Delivering Integrated Flood Risk Management

Governance for collaboration, learning and adaptation

Cover:
Signatory event for the dyke relocation project in Nijmegen, along the river Waal.
Courtesy of Room for the River

Delivering Integrated Flood Risk Management

Governance for collaboration, learning and adaptation

Dissertation

Submitted in fulfillment of the requirements of
the Board for Doctorates of Delft University of Technology
and of the Academic Board of the UNESCO-IHE
Institute for Water Education
for the Degree of DOCTOR
to be defended in public on
Friday, 14 February 2014 at 10:00 hours
in Delft, The Netherlands

by

Sebastiaan VAN HERK
Master of Science in Systems Engineering, Policy Analysis and Management
born in Lekkerkerk, the Netherlands

This dissertation has been approved by the supervisors:

Prof. dr. ir. C. Zevenbergen
Prof. R.M. Ashley

Composition of Doctoral Committee:

Chairman	Rector Magnificus TU Delft
Vice-Chairman	Rector UNESCO-IHE
Prof. dr. ir. C. Zevenbergen	UNSECO-IHE / TU Delft, supervisor
Prof. R.M. Ashley	UNESCO-IHE / Univ. of Sheffield, supervisor
Prof. mr. dr. E.F. ten Heuvelhof	TU Delft
Prof. dr. ir. A. E. Mynett	UNESCO-IHE / TU Delft
Prof. dr. ing. S. Schaap	TU Delft
Ir. I. de Boer	Rijkswaterstaat, Utrecht
Prof. dr. ir. M. J. C. M. Hertogh	TU Delft, reserve member

CRC Press/Balkema is an imprint of the Taylor & Francis Group, an informa business

Published by:
CRC Press/Balkema
PO Box 11320, 2301 EH Leiden, The Netherlands
e-mail: Pub.NL@taylorandfrancis.com
www.crcpress.com – www.taylorandfrancis.com

ISBN 978-1-138-02632-2 (Taylor & Francis Group)

Contents

Figures and tables

List of figures

List of tables

Preface

This dissertation combines the outcomes of multiple research projects. These projects have resulted in several interrelated papers of which six are included in this thesis. The presented thesis should, therefore, be considered a thesis by papers. However, from the outset of the research, these papers were written to an overall plan to create a coherent story. This plan was driven by the ambition to assist professionals in delivering integrated flood risk management in practice and to bring new insights to various related scientific domains. This thesis is the result of that plan that was also adapted to opportunities that came on to my path and has evolved and deepened as I progressed.

Sebastiaan van Herk

30 December 2013

This research has been co-funded by:

The Interreg IVB North Sea Region Programme

Investing in the future by working together for a sustainable and competitive region

leven met water

ruimte voor de rivier

Summary

Guidance for Integrated Flood Risk Management

The frequency and consequences of extreme flood events have increased rapidly worldwide in recent decades, and climate change and economic growth are likely to exacerbate this trend. Flood protection measures alone cannot accommodate the future frequencies and impacts of flooding. Integrated flood risk management (IFRM) considers a portfolio of measures to manage flood risk that comprises flood protection, but also land use planning, emergency management and other measures.

The implementation of IFRM policies and projects is not straightforward and is hampered by multiple governance challenges. IFRM requires collaboration between many disciplines and by a group of stakeholders with various and often diverse interests and means. IFRM requires the combination of objectives and funding from different policy domains; and to consider a range of possible options at all spatial scale levels and for various time horizons.

To date, no exhaustive, evidence-based guidance to deliver IFRM projects has been produced. In addition, such projects need to be delivered within the overarching context of an incumbent societal system comprising institutions and policies that are not (yet) designed for IFRM. By their implementation, IFRM projects will not only deliver on their objectives (outputs) and change the physical system (natural and man-made), but would also generate outcomes that have an impact beyond the scope of the project and that are sustained after the delivery of the project. Projects can contribute to a transition that changes the structures, cultures and practices of the societal system to enable a more widespread implementation of IFRM.

The objective of this research is to provide guidance to organise projects that aim to deliver IFRM, and, in doing so, to support a transition from the traditional flood protection regime to one of IFRM.

Research approach: 4 case studies and multiple methods

The research presented in this thesis has examined four case studies in the Netherlands that have all adopted an IFRM approach, but have each developed different options to reduce flood risk.

- Stadswerven: a flood proof urban development of a neighbourhood in the outer marches of Dordrecht.

- Westflank: creation of water storage capacity in a regional urban development in Haarlemmermeer.
- Island of Dordrecht: used the multi-layered-safety approach to combine flood protection, spatial planning and emergency management to increase the flood safety of the polder area protected by a ring dyke.
- Room for the River (RftR): delivered river widening measures to increase the river discharge capacity of the rivers Rhine, Meuse, Waal, IJssel and Lek.

The case studies have been examined for the organisation of the collaborative planning process, the delivery of integrated outputs and the generation of outcomes. Several analytical and theoretical frameworks have been developed from existing theories, data analysis and case study comparison. The research includes multiple methods: action research; document analysis; semi-structured and group interviews (205 interviewees); surveys (193 survey respondents); validation interviews and workshops (550 interviewees and participants); and observation.

The role of the author of this thesis has been different for each case study, ranging from project manager and action researcher to external researcher and observer. This has allowed for a mix of development and validation of new theories and for contributing to practice as well as science. This research led to 21 scientific publications with contributions from the author. Six of these are included in this thesis.

IFRM outputs: integration of objectives and across spatial and temporal scales

The objectives of IFRM have broadened from reducing the probability of flooding to reducing and managing flood risk, which also includes reducing potential consequences of flooding. This necessitates a balance with other spatial and land use planning objectives such as housing, nature, economic development and transport. Balancing multiple objectives inherently requires integration across spatial scales. IFRM implies a different framing of problems and flood risk objectives at various spatial scale levels. Options and measures at all spatial scale levels (e.g. from individuals to communities, and from individual buildings and neighbourhoods to an entire city or catchment) are to simultaneously: protect; prevent; and/or prepare against/for flooding. Moreover, the effectiveness of a given measure depends on integration with other measures implemented at other spatial scales and levels that may be outside the scope of the particular project.

This research shows that balancing multiple objectives inherently requires integration across temporal scales. Traditionally flood management has been based on a regulatory approach. Design standards are set and regulated for dykes and drainage systems to withstand up to specific return periods of rain or flow events or water levels in the rivers. In contrast, scenario planning that comprises inter alia the analysis of more extreme flood events than anticipated by the design standard has shown that the IFRM concepts as applied in the case studies, are more robust or provide flexibility to deal with climate change and related uncertainties regarding return periods, water levels and flows. Combining or selecting measures from as wide a range of options as possible requires balancing short and long-term costs and benefits and anticipating (potential) future change and uncertainties. Different options, such as flood protection systems, urban developments and emergency plans have different economic and technical lifespans, and their planning, operation and maintenance processes differ.

Guidance for planning processes to deliver IFRM outputs

This research provides guidance to organise the planning process of IFRM projects to deliver integration of objectives across spatial and temporal scales.

A project should start with multiple objectives and an integrative vision or concept such as river widening (RftR), flood proofing (Stadswerven), building with water storage (Westflank), or multi-layer-safety (Island of Dordrecht). The project scope should be granted design freedom to devise new measures and combinations thereof. Governance arrangements should stimulate collaboration and public participation, as the knowledge and means of all stakeholders are necessary. Moreover the collaborative planning process is to establish legitimacy, build trust and embed accountability systems to safeguard support for integrated processes and outputs throughout the planning process. The collaboration structure is to be fit for purpose and adaptable, whilst balancing top-down and bottom-up governance, and formal and informal relations. Finally, the planning process needs to stimulate learning between stakeholders and disciplines, and embed feedback loops to absorb threats and seize opportunities from economic, political and legislative dynamics.

The frameworks presented in Figure a and Figure b have been developed to structure and analyse collaborative planning processes in IFRM projects. The framework presented in Figure a organises project activities (that can comprise different work tasks) along three lines: 1. system analysis; 2. collaborative planning, design and engineering; and 3. governance. These activities generate knowledge (establish facts; create images and set ambitions) and support decision-making (address problems, develop solutions, involve participants, and influence politics). The activities are to be mutually enriching. They undergo a process of divergence (knowledge development in one activity) and convergence (knowledge exchange between activities) in each planning phase. The mutual enrichments of proposing (designing), analysing and selecting (governance) options contributes to the integrated nature of the outputs.

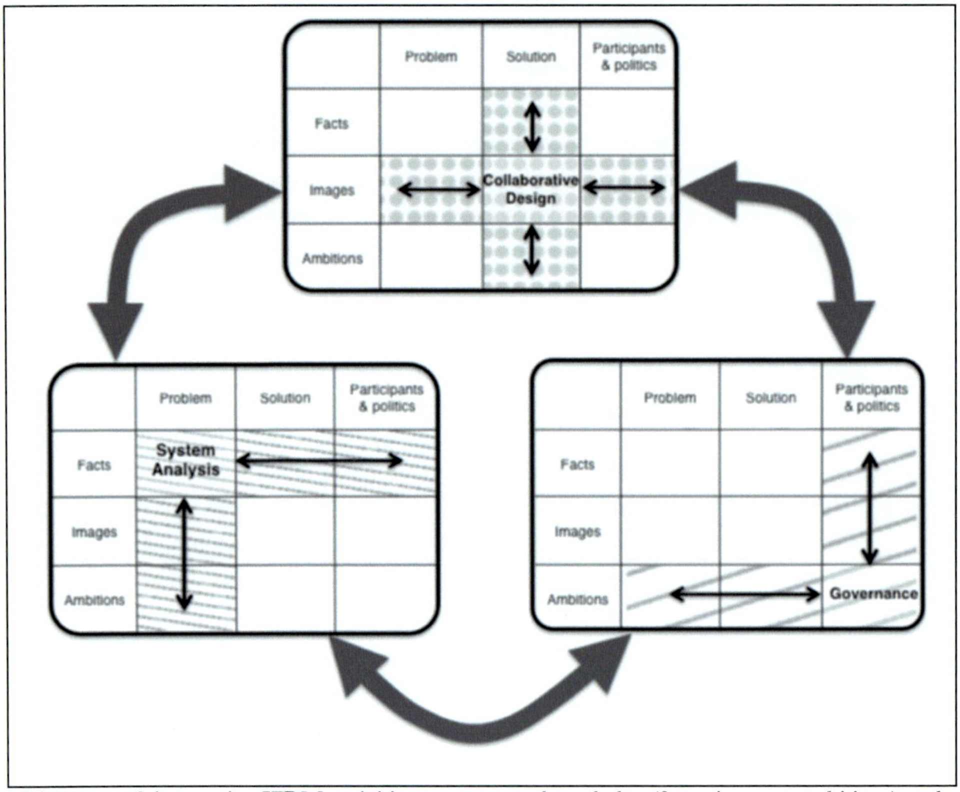

Figure a, 3 interactive IFRM activities to generate knowledge (facts, images, ambitions) and support decision making (problem, solution, participants & politics). Van Herk et al., 2011a.

The activities can be analysed or (re)designed successively for different planning phases (Figure b). In the first phase (phase 1) of 'divergence': the system performance is analysed, e.g. in terms of the probability and potential consequences of flood events; stakeholders are brought together and together set their objectives; and different strategies, options or measures to intervene in the flood risk management system are explored.

Bringing the outcomes of these series of activities together at the interfaces between activity types leads to a new phase (phase 2). Stakeholders can (re)frame problems and engage in joint goal-setting and discuss strategies and options based on the combined means and objectives of all stakeholders. The performance of identified strategies and options can be assessed, also using the objectives discussed.

Ultimately (phase 3) the options are to be combined into an IFRM plan comprising a portfolio of measures and an investment and implementation plan that is subject to formal decision-making processes between the stakeholders and within the democratically representative bodies of governmental organisations involved. Also, the performance of the IFRM plan is to be analysed, which inherently is a feedback loop to the continuous monitoring of system performance (phase 1) that will change with the implementation of the IFRM plan and due to exogenous factors.

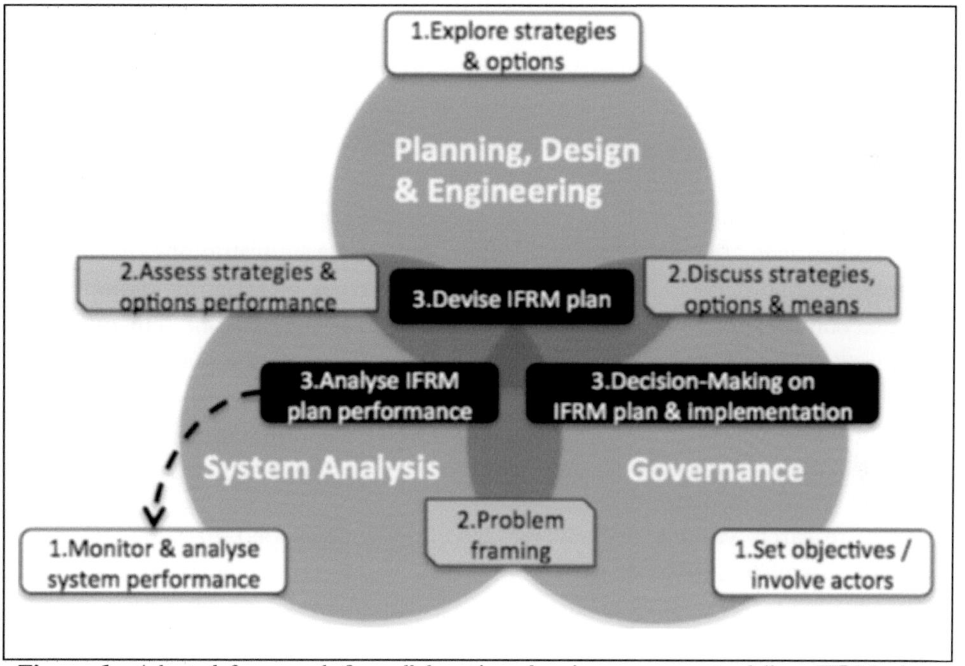

Figure b, Adapted framework for collaborative planning processes to deliver IFRM plans focusing on interfaces between activities in 3 phases. Van Herk et al., 2013a.

Project outcomes to contribute to a transition to IFRM

The projects researched in this thesis have taken place during, and have contributed to, a transition to IFRM. To date, literature in project management and transition management have not (yet) considered how individual projects can contribute to a transition. In practice, a focus on project outcomes and the dissemination of lessons is not self-evident either.

The projects have contributed to the transition by generating outcomes beyond the scope of the project and that are sustained after the duration of the project. New options (such as river widening and flood proofing) have been demonstrated and related practices and methods for the design and analysis of these options have been developed. These can be replicated elsewhere. Policy and regulation have been adjusted to allow for the implementation of new options in the case studies and for future projects. New governance arrangements (such as a programme directorate) have been developed in the case studies and are currently being used to organise new projects and programmes such as the Dutch Delta Programme and Flood Protection Programme. The projects described in the thesis contributed to capacity building and creation of networks amongst individuals and organisations that have been involved in the case studies. They have indicated that this will support them to implement future IFRM policies and projects.

This research draws lessons on governance arrangements that are to stimulate the delivery of projects within their contexts, and also to stimulate the generation of outcomes to change the regime. Projects and programmes are to: be managed at the relevant biophysical scale levels (e.g. river basin); be organised for polycentric governance (e.g. between national and regional administrations in RftR); stimulate public participation; comprise different types of experimentation; and create a culture and network to stimulate learning and share lessons. The different collaborative frameworks that have been used (collaborative research, Learning and Action Alliances (LAAs), or an embedded institution such as RftR's programme directorate) stimulated learning and were continuously adapted based on lessons learnt during the projects.

In this research various frameworks have been developed to stimulate and evaluate learning. An evaluation framework based on network learning can support a structured analysis of various learning outcomes of projects. The framework classifies learning outcomes into 3 categories: interpretations (philosophies or paradigms), structures (adjust governance arrangements and patterns), and practices (cognitive and behavioural).

The use of this typology has enabled the analysis of interdependencies and feedback loops between the learning outcomes. The framework merits further validation because it has shown potential to analyse simultaneously the contribution of projects to: changes to the physical system (outputs); to the societal systems (outcomes); and, in addition, to adjustments of governance arrangements of projects themselves (processes and structures) – all of which have been relevant to deliver IFRM in the case studies.

The results of this research indicate that the conceptual boundaries between adaptive co-management and transition management become opaque for projects. The learning and collaboration that takes place in projects is instrumental to changes to both the physical system and to the societal system, and the learning outcomes are generated simultaneously.

Samenvatting

Richtlijnen voor integraal (hoog)waterbeheer

De frequentie en consequenties van extreme overstromingen zijn wereldwijd snel toegenomen in de afgelopen decennia. Klimaatverandering en economische groei zullen deze trend waarschijnlijk verergeren. Louter maatregelen ter bescherming tegen hoogwater, zoals dijken, zijn op termijn naar verwachting onvoldoende om de risico's van overstromingen te kunnen beheersen. Integraal (hoog)waterbeheer (IFRM) beschouwt een portfolio van maatregelen voor de beheersing van overstromingsrisico's en wateroverlast. Dit portfolio omvat beschermingsmaatregelen, maar ook ruimtelijke ordening en rampenbeheersing.

De implementatie van IFRM beleid en projecten is niet eenvoudig en wordt bemoeilijkt door meerdere bestuurskundige en beleidsmatige, ofwel *governance* opgaven. IFRM vraagt om samenwerking tussen meerdere disciplines en tussen een groep actoren met verschillende belangen en middelen. Dit is nodig om doelstellingen en financiering vanuit verschillende beleidsdomeinen te combineren, en/of om maatregelen op meerdere ruimtelijke schaalniveaus en voor meerdere tijdsschalen te verbinden.

Er zijn geen uitvoerige, wetenschappelijk onderbouwde richtlijnen gevonden voor het succesvol initiëren en uitvoeren van IFRM projecten. Een complicerende factor hierbij is dat voor dergelijke projecten een passende maatschappelijke en institutionele context (nog) ontbreekt. De implementatie van IFRM projecten moet niet alleen leiden tot de gewenste verandering van het fysieke systeem (c.q. halen van projectdoelstellingen (*outputs*)), maar ook uitkomsten genereren die buiten de projectscope en na de projectduur doorwerking zullen hebben (*outcomes*). Projecten kunnen zo bijdragen aan een transitie van het maatschappelijk systeem doordat de bestaande structuren, culturen en praktijk meeveranderen. Zo wordt een bredere implementatie van IFRM mogelijk.

De doelstelling van dit onderzoek is om richtlijnen te ontwikkelen voor de organisatie van projecten in integraal (hoog)waterbeheer, en zodoende ook om een bijdrage te leveren aan de transitie van overstromingsbescherming naar integraal (hoog)waterbeheer.

Onderzoeksaanpak: 4 case studies en meerdere methoden

Het onderzoek gepresenteerd in dit proefschrift heeft 4 case studies in Nederland bestudeerd die allen een IFRM aanpak hebben gebruikt, maar verschillende maatregelen hebben ontwikkeld om het overstromingsrisico te verlagen.

- Stadswerven, een hoogwaterbestendige stedelijke ontwikkeling in het buitendijkse gebied van Dordrecht.
- Westflank, het creëren van waterbergingscapaciteit voor de gebiedsontwikkeling in de Haarlemmermeer.
- Eiland van Dordrecht, waar de meerlaagsveiligheid benadering is gebruikt om de waterveiligheid te verhogen voor dijkring 22 door bescherming, ruimtelijke ordening en rampenbeheersing te combineren.
- Ruimte voor de Rivier (RvdR), de rivierwaterafvoer verhogen door rivierverruiming langs de rivieren Rijn, Maas, Waal, IJssel en Lek.

De case studies zijn bestudeerd op de organisatie van het planningsproces, de realisatie van integrale *outputs* en de generatie van *outcomes.* Diverse raamwerken zijn ontwikkeld uit bestaande theorieën, data analyse en case study vergelijking. Het onderzoek omvat meerdere methoden: actie onderzoek, documenten analyse, semi-gestructureerde- en groepsinterviews (205 interviews), vragenlijsten (193 respondenten), validatie interviews, validatie workshops (550 geïnterviewde personen en deelnemers) en observatie. De rol van de auteur van dit proefschrift is verschillend geweest voor iedere case studie. Hij is o.a. project manager en actie onderzoeker geweest, als ook externe onderzoeker en waarnemer. Hierdoor is het mogelijk geweest om nieuwe theorieën te ontwikkelen en te valideren, alsmede een bijdrage te leveren aan zowel de praktijk als aan de wetenschap. Het onderzoek heeft geleid tot 21 wetenschappelijke publicaties met bijdragen van de auteur. Hiervan zijn 6 publicaties opgenomen in dit proefschrift.

IFRM outputs: integratie van doelstellingen, ruimtelijke- en tijdschalen

De doelstellingen van IFRM zijn verbreed van het verlagen van de kans op overstromingen naar het verlagen van de overstromingsrisico's. IFRM omvat dus ook de verlaging van de mogelijke consequenties van overstromingen. Hierdoor is het noodzakelijk om ook andere doelstellingen in het domein van de ruimtelijke ordening en planning zoals huisvesting, natuurbeheer, economische ontwikkeling en mobiliteit in de afweging mee te nemen. Dit onderzoek laat zien dat bij het afwegen van meerdere doelstellingen integratie tussen ruimtelijke schaalniveaus onvermijdelijk is. IFRM impliceert een aangepaste probleemdefinitie en andere doelen op verschillende ruimtelijke schaalniveaus. Op diverse ruimtelijke schalen in het fysieke systeem (van individuen tot gemeenschappen, en van individuele gebouwen en wijken tot een gehele stad of stroomgebied) zijn er gelijktijdig maatregelen nodig ter bescherming tegen, preventie van, en voorbereiding op overstromingen. Bovendien hangt de effectiviteit van een bepaalde maatregel af van maatregelen die genomen worden op andere ruimtelijke schaalniveaus die vaak buiten de directe doelstelling van het project liggen.

Uit dit onderzoek blijkt dat het afwegen van meerdere doelstellingen ook de integratie tussen meerdere tijdschalen behoeft. Traditioneel waterbeheer is gebaseerd op de regulering van ontwerpstandaarden voor dijken en drainage systemen om bepaalde rivierwaterstanden respectievelijk regenbuien tot bijbehorende herhalingsperioden te kunnen weerstaan. Echter, IFRM concepten zoals toegepast in de case studies hebben de potentie om veerkrachtiger te zijn om extremere overstromingen dan de ontwerpstandaard te doorstaan. Zij kunnen ook flexibiliteit bieden om in te spelen op klimaatverandering en gerelateerde onzekerheden omtrent de herhalingsperioden van waterstanden en regenbuien. Het combineren en selecteren van maatregelen behoeft het afwegen van korte en lange termijn kosten en baten en het anticiperen van mogelijke toekomstige veranderingen en onzekerheden. Maatregelen zoals dijken, stedelijke ontwikkeling of rampenplannen hebben een verschillende economische en technische levensduur. Dit geldt dus ook voor hun plannings-, beheer- en onderhoudsprocessen.

Richtlijnen voor planningprocessen met IFRM outputs

In dit onderzoek zijn richtlijnen ontwikkeld voor het planningsproces van IFRM projecten om te komen tot integratie van doelstellingen, ruimtelijke- en tijdschalen.

Een project moet starten met meerdere doelstellingen en een integraal concept zoals rivierverruiming (RvdR), waterbestendig bouwen (Stadswerven), bouwen met waterberging (Westflank) en meerlaagsveiligheid (Eiland van Dordrecht). Het project moet een bepaalde ontwerpvrijheid krijgen om nieuwe (combinaties van) maatregelen te kunnen ontwikkelen. Organisatiestructuren en -processen moeten participatie van, en samenwerking tussen verschillende actoren stimuleren. Hun gezamenlijke kennis en middelen zijn nodig. Bovendien leidt dit tot legitimering van, en vertrouwen in het planningsproces. Tot slot moet het leren tussen actoren en disciplines gestimuleerd worden, ook om te kunnen omgaan met de politieke en economische dynamiek.

De raamwerken uit Figuur a en Figuur b helpen het planningsproces te structuren (vooraf) en analyseren (achteraf). Het raamwerk uit Figuur a structureert project activiteiten langs drie lijnen: 1. systeem analyse, 2. planning, ontwerp en techniek, en 3. *governance*. Deze activiteiten genereren kennis (feiten-, beelden- en wilsvorming) en ondersteunen besluitvorming (adresseren van problemen, ontwikkelen van oplossingen, en het betrekken van actoren en beïnvloeden van politiek). De activiteiten verrijken elkaar. Zij gaan door een proces van divergentie (kennisontwikkeling binnen een activiteit) en convergentie (kennisuitwisseling tussen activiteiten) in iedere opvolgende planningsfase. De onderlinge verrijking van het ontwerpen, analyseren en selecteren draagt bij aan de integrale aard van de *outputs*.

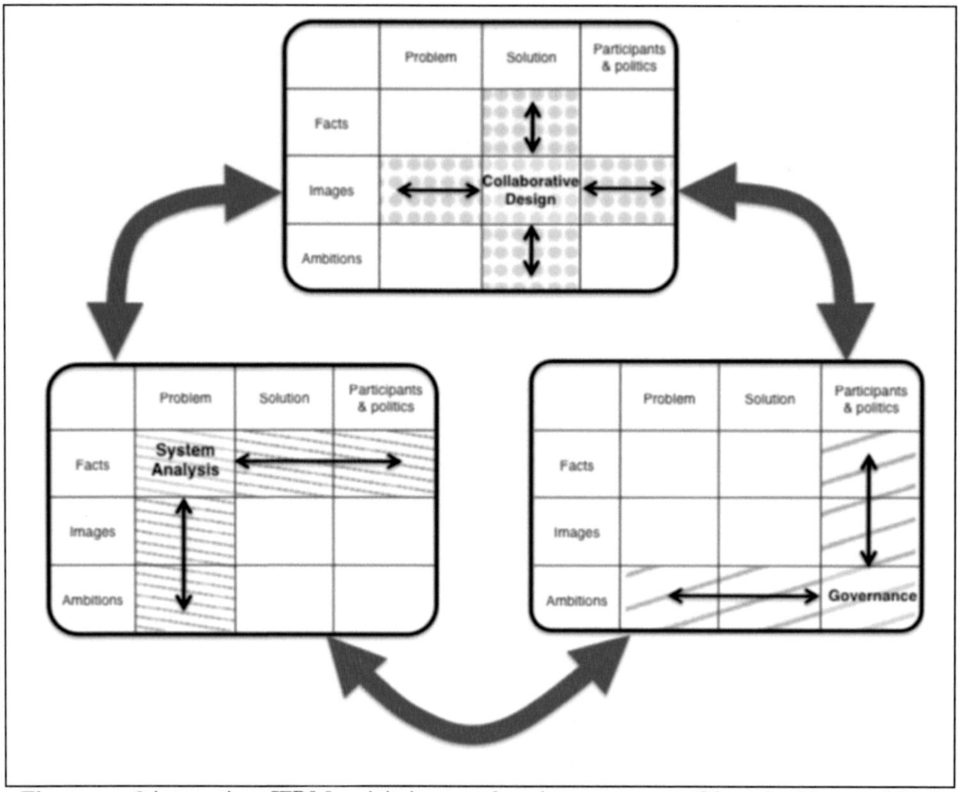

Figuur a, 3 interactieve IFRM activiteiten om kennis te genereren (feiten, beelden, ambities) en besluitvorming te ondersteunen (problemen, oplossingen, participanten & politiek). Van Herk et al., 2011a.

De activiteiten kunnen geanalyseerd of (her)ontworpen worden voor verschillende opvolgende planningsfasen (Figuur b). In de eerste fase (fase 1) van 'divergentie' wordt het functioneren van het systeem geanalyseerd, bijvoorbeeld op de kans op, en mogelijke consequenties van overstromingen. Actoren worden bij elkaar gebracht om doelstellingen te bepalen en strategieën of maatregelen te verkennen. Het verbinden van de uitkomsten van deze activiteiten leidt tot een nieuwe fase (fase 2). Op basis van de inzichten in het functioneren van het systeem en de verschillende doelstellingen kunnen actoren de opgaven (her)definiëren. Bij de beoordeling van de strategieën en maatregelen spelen de belangen en gezamenlijke middelen van alle actoren een rol. Tot slot (fase 3) worden de maatregelen gecombineerd in een integraal waterbeheer (IFRM) plan. Dit plan omvat behalve een portfolio aan maatregelen ook een investering- en implementatieplan. Het toekomstig functioneren van het watersysteem na de implementatie van het IFRM plan wordt ook geanalyseerd, wat inherent weer een feedback lus is naar het continue monitoren van het functioneren van het systeem (fase 1).

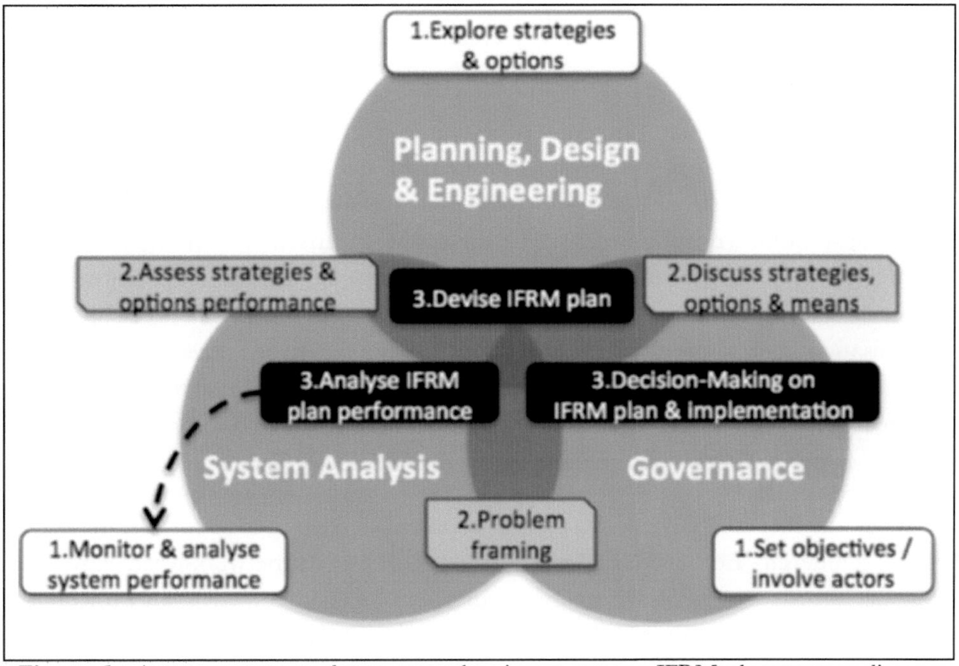

Figuur b, Aangepast raamwerk voor een planning proces om IFRM plannen te realiseren, met bijzondere nadruk op de overlap tussen activiteiten. Van Herk et al., 2013a.

Project *outcomes* voor bijdrage aan transitie naar IFRM

De projecten onderzocht in dit proefschrift hebben plaats gevonden tijdens, en hebben bijgedragen aan de transitie naar integraal waterbeheer (IFRM). Project management en transitie management literatuur beschouwden tot op heden (nog) niet hoe individuele projecten kunnen bijdragen aan een transitie. In de praktijk is aandacht voor de doorwerking van kennis en ervaring ook niet vanzelfsprekend.

De projecten hebben bijgedragen aan een transitie door *outcomes* te genereren die buiten de projectscope en na de projectduur hun doorwerking hebben gehad. Nieuwe maatregelen (zoals rivierverruiming en waterbestendig bouwen) zijn gedemonstreerd en nieuwe methoden voor het ontwerp en analyse van deze maatregelen zijn ontwikkeld. Deze kunnen elders ook toegepast worden. Beleid en wetgeving (zoals voor buitendijkse ontwikkelingen) zijn aangepast om de nieuwe maatregelen te kunnen implementeren in de case studies én in toekomstige projecten. Nieuwe organisatiestructuren en –processen (zoals een programma directie of ontwerpend onderzoek) zijn ontwikkeld en worden inmiddels gebruikt in nieuwe projecten en programma's zoals het Delta Programma en het HoogWater BeschermingsProgramma. De projecten hebben bijgedragen aan het leren van individuen en organisaties en het ontstaan van netwerken tussen hen. De respondenten van dit onderzoek hebben aangegeven dat dit hen zal ondersteunen om toekomstig beleid en projecten te implementeren.

Dit onderzoek wijst op diverse strategieën om projecten zo te organiseren, dat deze gerealiseerd worden binnen een bestaande context (*outputs*), én ook *outcomes* genereren die bijdragen aan een transitie van de context. Projecten moeten georganiseerd worden op het juiste ruimtelijke schaalniveau (bijv. stroomgebied), meerlaagse besluitvorming faciliteren (bijvoorbeeld tussen rijk en regio zoals in RvdR) en publieke participatie organiseren. Projecten moeten een cultuur en netwerk voor leren organiseren, met experimenteerruimte. De verschillende organisatiemodellen die gebruikt zijn (ontwerpend onderzoek, *Learning and Action Alliances* (LAAs), of een programma organisatie zoals voor Ruimte voor de Rivier) stimuleerden het leren. De organisatiestructuren zijn bovendien continu aangepast op basis van de leerervaringen gedurende de projecten.

In dit onderzoek zijn diverse raamwerken ontwikkeld om het leren te stimuleren en te evalueren. Een evaluatie raamwerk gebaseerd op netwerk-leren kan een gestructureerde analyse ondersteunen van diverse leer-uitkomsten van projecten. Het raamwerk classificeert leer-uitkomsten in 3 categorieën; interpretaties (filosofieën of paradigma's), structuren (organisatiemodellen en patronen) en praktijken (cognitief en gedrag). Het gebruik van deze typologie heeft de analyse mogelijk gemaakt van wederkerigheden en feedback lussen tussen de leer-uitkomsten. Het raamwerk kan gelijktijdig projecten analyseren op hun bijdrage aan: veranderingen aan het fysieke systeem (*outputs*), de maatschappelijke systemen (*outcomes*), en aanpassingen van de organisatiemodellen van de projecten zelf (processen en structuren). Deze bijdragen zijn allemaal relevant gebleken om IFRM projecten te realiseren in de case studies.

De resultaten van dit onderzoek wijzen erop dat de conceptuele grenzen tussen adaptief co-management en transitie management vervagen op het niveau van projecten. Het leren en de samenwerking die plaats vindt in projecten is instrumenteel voor de veranderingen van het fysieke systeem en de transitie, en de leer-uitkomsten worden gelijktijdig gegenereerd.

Chapter 1

Introduction

Chapter 1 Introduction

1.1 Background and problem

The frequency and consequences of extreme flood events have increased rapidly worldwide in recent decades (e.g. Bouwer et al., 2007; Kron, 2009; Munich Re, 2009) and climate change is likely to exacerbate this trend in the near future (e.g. IPCC, 2007). The key factors for this increase in flood risk are: climate variability and extremes (UNISDR, 2011a; 2012); global population growth and the increase in socio-economic activities in flood prone areas (*ibid*), together with their growing interdependency on flood protection and drainage infrastructure of which a significant part is of unknown or poor condition (Ashley and Cashman, 2006; National Committee on Levee Safety, 2009).

1.1.1 Towards an integrated approach to flood risk management (IFRM)

Traditionally, flood management practices in Europe have focused on predominantly hazard control, or i.e. flood protection measures such as dykes or drainage systems to reduce the probability of flooding (Klijn et al., 2008; Newman et al, 2011). However, in the past two decades major flood disasters have created the need to shift from flood protection to a more integrated approach in which flood risk is actively managed to also reduce flood impacts (White, 2010; Dawson et al. 2011). It is increasingly recognised that engineering responses alone cannot accommodate the future frequencies and impacts of flooding (Yovel, 2013; UNISDR, 2012; 2011a; 2011b). Moreover, the mere use of large infrastructure, particularly flood protection, has the risk for 'technological lock-in' (Walker, 2000) or for 'investment trap' (Belt et al., 2013), creating a path dependency that reduces the opportunities to take alternative or complementary measures. In addition to the development, implementation and operation of flood protection measures, integrated flood risk management (IFRM) considers additional measures to reduce flood risk. The portfolio of IFRM measures comprises: hard structural measures that aim to reduce risks by modifying the system through physical and built interventions; soft structural measures that involve maintaining or restoring the natural processes with the aim of reducing risks; or non structural measures that may not require engineering, but its contribution to risk reduction is often through changing behaviour through regulation, encouragement and/or economic incentivisation (Gersonius, 2012).

Multiple authors advocate IFRM to reduce flood risk (Zevenbergen et al., 2008; Huntjens et al., 2011; Pahl-Wostl et al., 2012). Apart from flood protection measures, particularly land use planning is considered as one of the more crucial components in managing flood risk. It can reduce exposure and vulnerability to flooding by, inter alia, discouraging developments in flood prone areas and stimulating the flood proofing of developments (Yovel 2013; Wheater & Evans, 2009; White, 2010).

IFRM is a central concept to new policies such as: the EU Flood Directive (EC, 2006); the source-pathway-receptor framework as used by the Environment Agency in England and Wales (EA, 2000); and the multi-layer safety approach (MLS) in the Netherlands (V&W, 2008). The EU Flood Directive requires the EU member states to prepare IFRM plans for river basin and coastal areas at risk of flooding (although it omits urban drainage and smaller streams and tributaries). The IFRM plans should include appropriate objectives and measures that focus on the reduction of the likelihood of flooding and/or on the reduction of the potential adverse consequences. IFRM plans are meant to address all phases of the flood risk management cycle, particularly: prevention; protection; and preparedness (adapted from: EC, 2004).

In certain countries, such as England, planning incorporating flood risk has become a mandatory requirement (since 2001) and formerly took a regional as well as a local perspective (DCLG, 2009). However, this has now been replaced by a less prescriptive and more open planning framework (DCLG, 2012b). In the Netherlands, the National Water Plan (V&W, 2008) proposes the three-layered MLS approach that goes beyond flood protection, which is its first safety layer. The two other layers are aimed at reducing the consequences of flooding by adapting the spatial layout (second safety layer) and enhancing emergency response (third safety layer), respectively. Hence, MLS explicitly calls for the consideration of measures to reduce adverse consequences and to operationalise these measures. The Dutch spatial planning procedure 'Room for the River' comprises river widening measures and explicitly aims to increase flood safety combined with increased spatial quality of landscape, nature and culture (Ministry of Transport, Public Works and Water Management, 2000; Schut et al., 2010). In addition to planning for new developments, the large stock of existing buildings provides significant opportunities for retrofitting as part of normal redevelopments. Thus it may be possible to incrementally reduce flood vulnerability in European cities by taking advantage of current redevelopment opportunities. For this to happen, flood risk management needs to be integrated better into planning and urban retrofitting and development processes (ECTP,

2005).

The implementation of IFRM policies and delivery of a portfolio of IFRM measures faces multiple additional governance challenges compared to traditional flood management (Section 1.1.2). This research addresses these challenges and provides guidance, which has been lacking to date (Section 1.1.3), to organise projects[1] that aim to deliver IFRM.

1.1.2 Governance challenges for IFRM

Implementation of IFRM is not straightforward. The 'integrationist agenda' as described by Medema et al. (2008) gives an idea of how broad its scope can be and implicitly shows the inherent complexity related to organising for integration: "the integrated and coordinated management of water and land as a means of balancing resource protection while simultaneously meeting social and ecological needs and promoting economic development". IFRM can be considered a fundamental component of integrated water management (IWM) (Green, 2010). The theoretical development of these domains has been mutually supportive up to the point that definitions of IFRM and IWM are interchanged (e.g. Wolsink, 2006). The definition of Thomas and Durham (2003) for IWM is directly applicable to IFRM: "..an approach to water <flood> management that recognises its multidimensional character—time, space, multidiscipline and stakeholders—and the necessity to address, embrace and relate these dimensions holistically....". These multiple dimensions of IFRM bring about many governance challenges to organise for the implementation of IFRM policies and projects. In this Section (1.1.2) an introduction is given to the governance challenges. In the conclusions of the research presented in this thesis (Chapter 8.1) they are structured to frame the guidance that this research provides to deliver IFRM.

The objectives of IFRM have broadened with respect to traditional flood management from reducing the probability of flooding, to reducing flood risk, which also includes reducing potential consequences of flooding. Flood risk inherently comprises multiple and sometimes competing objectives, such as reducing: individual risk (probability of death of one person due to flooding); group risk (function of probability of flood event and related number of fatalities); economic risk (function of probability of flood and related direct and indirect material damage. Indirect adverse economic consequences can also include

[1] A project is defined here as a temporary endeavour with a defined beginning and end (usually

foregone revenues). Moreover, physical interventions to reduce flood risk need to be incorporated into spatial planning and thus need to be balanced with a range of other spatial planning priorities and objectives, including meeting new housing needs, facilitating economic growth, and creating and maintaining quality places (Potter et al., 2013) to increase the political and financial feasibility of the implementation of measures that reduce flood exposure and vulnerability (White, 2010; Veerbeek et al, 2012). Spatial planning sets out to integrate various needs and requirements at a range of spatial and temporal scales. In this, flood risk management is not normally considered to be the most important of the various utility and service needs and opportunities such as e.g. mobility, or energy supply. Contrary, the implementation of flood protection measures often prevails over other planning objectives when they need to be integrated in spatial or zoning plans (White, 2010; Klijn et al., 2008).

There is a lack of understanding or shared perception of the effectiveness of IFRM measures (Adger et al., 2005) such as, inter alia, urban planning or emergency planning. This holds particularly true for a portfolio of measures that is to be implemented gradually as part of an adaptation strategy (*ibid*; Gersonius, 2012). This lack of understanding contributes to unwillingness by involved professionals to take risk and innovate by using alternative measures to flood protection (Newman et al., 2011). The flood management discipline is dominated by engineers ¨who tend to look for solutions in a direction familiar to them¨ (Klijn et al., 2008) and thus sustain a culture of flood protection. Partnerships and collaboration are essential to achieve an integrated approach. Planners need the expertise, and crucially the understanding, of engineers and hydrologists. However, there can be considerable misunderstanding and miscommunication between disciplines (Potter et al., 2013; Zhou et al, 2013).

Flood risk reduction is a long-term goal compared with many other planning considerations such as increasing mobility or facilitating economic growth. Balancing planning objectives for various time horizons further complicates the integration of flood risk management in planning processes (Gersonius, 2012). The implementation through spatial planning of a portfolio of measures to reduce flood risk requires a long-term plan or adaptation strategy that comprises multiple interventions over time (ibid). Some interventions might be needed in the short term to deliver the long-term plans and they require short-term benefits to be politically appealing (Geldof, 2007; Hamin & Gurran, 2009; White, 2010). The performance of flood risk reducing measures is to be evaluated for various time horizons and over a longer period of time to be able to fully appreciate their effectiveness (Carter et al, 2005; Belt et al., 2013). The increasing uncertainty about changing flood risk due to climate change (Milly et al., 2008) makes the performance assessment of measures over various time horizons more complex.

The multi-objective planning process is complex and means that no single stakeholder group has final or absolute control over urban or spatial developments (Hajer and Zonneveld, 2000; Sellers, 2002). This multi-actor setting further complicates the way in which flood risk can be adequately addressed in planning processes (OECD, 2010). Planning should be flexible and dynamic enough to address complex challenges - such as flood risk - combining spatial quality with democratic legitimacy. Governance and network theories (Hanf and Scharpf 1978; Kickert et al.,1997; Marsh and Rhodes, 1992; Rhodes 1997; Scharpf 1997) indicate that multiple stakeholders, private and public, at various levels of government and from various policy domains, are becoming more actively involved in decision making to share interests, aims and ambitions and to develop a joint definition of the problems and potential responses. Interactive decision making through dialogue between various stakeholders is expected to result in richer policy proposals that can be implemented more efficiently and thus raise the democratic legitimacy of the decisions (Edelenbos & Klijn, 2006). There is no blueprint to organise for participation and interactive decision making; rather it requires customised and adaptive process management (*ibid*).

The governance challenges for organising the implementation of IFRM can be defined as 'wicked-problems'; being problems that have multiple and conflicting criteria for defining solutions, solutions that create problems for others, and no rules for determining when problems can be said to be solved (Rittel and Webber, 1973), and 'persistent problems' that are: complex; ill-structured; involve many stakeholders; are surrounded by structural uncertainties; and are hard to manage" (Rotmans, 2005). The wicked and persistent problems are to be overcome in each project that aims to deliver IFRM and change the physical system (natural and man-made). Additionally, many authors (e.g. Wostl et al., 2010; Bos and Brown, 2012) argue that a change of the societal system[2] is necessary to enable IFRM. The research presented in this thesis distinguishes between guidance to organise projects that comprise interventions to adapt the physical system (Section 1.1.3) and understanding how these projects can contribute to a transition that changes the societal system and enables a wider uptake and implementation of IFRM (Section 1.1.4).

[2] A societal system is a part of society that can be attributed a functioning and functioning is the way a societal system meets a societal need. The functioning of societal systems can be described by its: structures, cultures and practices (e.g. Rotmans and Loorbach, 2006; Rotmans and Loorbach, 2009; Van Raak, 2010).

1.1.3 Lack of guidance to deliver IFRM projects

No exhaustive, evidence-based guidance has been found to organise for a multi-actor collaborative planning process to develop and implement IFRM plans comprising of various measures to obtain multiple objectives. The Global Water Partnership provides guidance (e.g. GWP-TAC, 2004) to implement IFRM, but Medema et al. (2008) show that the governance conditions and process steps proposed by the GWP are ¨not sufficiently detailed, are non-specific and untestable¨. Examples of such conditions and steps range from: having an enabling legislative and policy environment; appropriate institutional framework; to preparing a management strategy to be implemented and monitored. Logically, it is difficult to present conditions that are both generally applicable and sufficiently detailed, especially as the GWP focuses on policies and strategies and not on IFRM projects or programmes[3]. Moreover, few examples of IFRM projects have been identified, documented and evaluated (Pahl-Wostl et al., 2007) as the implementation of IFRM is still in its infancy (Huntjens et al., 2011) and evaluation frameworks for IFRM are not readily available.

Many authors stress the need for multi-actor collaboration and propose some sort of social learning framework (White, 2008; Boelens, 2006; Pahl-Wostl et al., 2007; Farrelly et al., 2009, etc.). Pahl-Wostl et al. (2007) recommend "social learning" processes as a means of developing and sustaining the capacity of different authorities, experts, interest groups, and the general public to manage their water systems effectively. Social learning includes both the capacity building of individuals and organisations, as well as the creation of relational qualities and social capital (Pahl-Wostl et al., 2007; Bouwen and Tallieu, 2004). Social learning also helps to build on past and continuing experience to cope with uncertainty and change, which is especially relevant for integrated flood risk management and urban planning (Folke, 2006). Social learning is considered an alternative, complementary policy instrument in water governance (Blackmore et al. (2007) and is gaining recognition as a potential governance or coordination mechanism (Ison & Watson, 2007). Ison et al. (2013) advocate not defining social learning rigidly, but rather leaving it to the practitioners to articulate and use the concept. All the abovementioned authors focus on social learning to manage the physical system or change the societal system, but do not address social learning at the level of projects for practitioners. In this thesis social learning is defined broadly and focused on its use in projects: the learning by individuals and organisations whilst working together on IFRM projects to change the physical system, whilst

[3] "a group of related projects managed in a coordinated way to obtain benefits and control not available from managing them individually" (Project Management Institute, 2008, p.434).

these social interactions are influenced by and may change the societal system. Few authors provide guidance on how to organise projects for social learning and they are not focused on supporting collaborative planning or flood risk management to deliver IFRM projects. The research presented in this thesis builds upon social learning theories to provide guidance to organise projects to deliver IFRM.

An increasing amount of literature in environmental management and integrated water management has emerged introducing the concepts of co-management, adaptive management, adaptive co-management and adaptive governance (e.g. Huitema et al., 2009; Armitage et al., 2008). The concepts address many elements related to the governance challenges of IFRM. They have various and partly overlapping definitions that comprise dimensions of learning and collaboration (*ibid*, Plummer and FitzGibbon, 2004; Berkes, 2009) to adapt to the complexity and uncertainty related to the physical or societal systems (e.g. Folke et al., 2005; Huitema et al., 2009). Various authors have developed conceptual frameworks for learning related to these concepts (e.g. Armitage et al., 2008; Steyaert and Jiggins, 2007; Pahl-Wostl, 2009; Folke et al, 2005, Huitema et al., 2009), but this literature, just as the social learning literature, does not provide guidance to organise for collaboration and learning in individual IFRM projects.

The research described in this thesis has taken 'projects' as the level of analysis that has been overlooked to date in this literature despite its significance to explain the adaptation of the physical and societal systems. The pro-active adaptation of the physical system takes place through interventions that are part of different projects that together form adaptation pathways (Gersonius, 2012). Learning and feedback loops that are necessary to adapt to complexity and uncertainty of the physical and societal systems (Folke et al., 2005) can occur within one project or between multiple projects. Projects or programmes present opportunities for social learning that can contribute to changes of the societal system (e.g. Armitage et al, 2008; Berkes, 2009; Rijke et al., in press). The research presented in this thesis aims to provide guidance to organise projects that aim to deliver IFRM. The guidance that this thesis will provide to organise projects, will explicitly address the role of learning in projects to adapt the physical and societal systems.

1.1.4 A transition to IFRM and the role of projects

Many scholars call for a transition of the societal system and change of its regime[4] from having a limited focus on flood protection: 'fighting against water', to actively managing flood risk that in addition to flood protection aims to reduce flood impacts and accommodate floods: 'living with water' (e.g.; White, 2010; Newman et al., 2011; Dawson et al. 2011; Zevenbergen et al., 2013b). Transition literature defines a transition as a fundamental change in the structures (the formal, physical, legal and economic aspects of functioning restricting and enabling practices), cultures (the cognitive, discursive, normative and ideological aspects of functioning involved in sense-making of practise) and practices (the routines, habits, formalisms, procedures and protocols by which actors, which can be individuals, organisations, companies, etc., maintain the functioning of the system) of a societal system, profoundly altering the way it functions (e.g. Rotmans and Loorbach, 2006; Rotmans and Loorbach, 2009; Van Raak, 2010).

The research presented in this thesis does not only provide guidance to deliver IFRM projects. In doing so it also addresses the governance challenges related to delivering a project in a regime that is yet 'unsympathic' to IFRM and provides understanding on the contribution of projects to a regime change to IFRM. A transition is needed to enable and facilitate a widespread uptake and delivery of IFRM projects (e.g. Pahl-Wostl, 2007; van der Brugge and Rotmans, 2007). The transition to IFRM is an already on-going process in various countries. In recent years, the implementation of policies such as the EU Floods Directive (EC, 2006) has stimulated the adaptation of the societal system in various EU countries by introducing new governance and institutional arrangements (Klijn et al., 2008). Governance experiments in Australia have stimulated learning amongst stakeholders that also contributes to changes of the societal system (Farrelly and Brown, 2011; Bos and Brown, 2012).

The transition to IFRM in the Netherlands, the geographical scope of the case studies in this thesis (Section 1.3.1), has started in the 1980s (Van der Brugge et al., 2005) and is still on-going (Zevenbergen et al., 2013b). In the Netherlands, traditional flood management focuses on hard structural flood protection measures and has clearly defined regulatory objectives and mandates and an aligned institutional context. To manage coastal and fluvial flood risk, the waterboards are responsible for operation and maintenance of the dykes, whilst the National Government sets related standards for dykes to withstand specific return periods of water levels and funds the investment costs of flood protection.

[4] The societal subsystem that dominates the functioning of the system (De Haan and Rotmans, 2011).

To manage pluvial flood risk in urban areas, the water boards operate the regional water system and water treatment plants, whilst municipalities operate the sewage systems and public space. IFRM plans can only be developed and implemented provided that these and other stakeholders are involved and that they can consider a wide range of possible options. Overarching policy that stimulates IFRM such as multi-level safety (V&W, 2008) or adaptive delta management (Deltacommissaris, 2011) is being developed in the Netherlands, but its implementation faces multiple barriers in the incumbent cultures, structures and practices of the societal system. The implementation of the multi-level-safety concept is hampered by current regulation and funding that favours flood protection measures (Ministry of Infrastructure & Environment, 2011) and by several experts that contend the effectiveness of the approach (WaterForum Online, 2012; Kolen and Kok, 2011). Approaches, methods and practices need to be developed or adjusted for adaptive delta management to address future uncertainties and develop adaptation pathways comprising a sequence of policy actions over time (Haasnoot, 2013). The integration of flood risk management into the way Dutch cities are planned, requires new skills and competencies in all of the stakeholders concerned (LMW, 2010).

Transition theories (e.g. Rotmans and Loorbach, 2006; De Haan and Rotmans, 2011) help describe how societal systems change, but do not focus on the role of projects during a transition. The hypothesis is posed here that projects that adopt an IFRM approach to deliver the project would also contribute to a regime change to IFRM. By their implementation, such projects will not only deliver on their objectives (outputs), but would also generate outcomes that have an impact beyond the scope of the project and that are sustained after the delivery of the project. Projects that take place during; are part of; and are a demonstration of a transition to IFRM, such as the case studies of this research (Section 1.3.1), need to be monitored and evaluated by practitioners and researchers alike. They are to document case studies and draw lessons on how projects can be delivered in an 'unsympathetic' regime and how projects can contribute to a transition by generating outcomes and organise for a wide uptake of the outcomes. This would provide guidance to practitioners on managing projects and would contribute to scientific theories on transitions. The research presented in this thesis aims to provide empirical evidence of how the transition from flood protection to IFRM has occurred in practice and analyse how projects that adopted an IFRM approach have contributed to the transition. It builds upon and enriches existing frameworks from transition literature (notably: De Haan and Rotmans, 2011) that to date have not been applied to analyse the impact of individual projects, nor to flood management. It concludes with lessons on governance arrangements that are to stimulate the delivery of projects within their present societal system, and that also stimulate the generation of outcomes to change the flood management regime.

1.2 Objective and research questions

The previous sections have described that if an integrated rather than traditional approach to flood risk management (IFRM) is adopted, more stakeholders and disciplines are to be involved to consider more measures and planning objectives in addition to flood protection at a variety of spatial and temporal scales. As a result the planning process becomes more complex. However, there is limited guidance on how to organize projects and programmes that aim to deliver IFRM to adapt the physical system. Additionally, the projects have to be delivered during, and can contribute to a transition of the societal system and its cultures, structures and practices to a IFRM regime that would enable a widespread uptake and implementation of IFRM projects. There is limited understanding and guidance on how projects contribute to regime change. Much scientific literature in flood management and transition management focuses either on managing and adjusting the physical system or the societal system. However, it overlooks the role of projects or combination of projects in contributing to the adaptation of both systems simultaneously. **This research provides guidance to organise projects that aim to deliver IFRM, and, in so doing, to support a transition from the traditional flood protection regime to one of IFRM.**

The main research question is:

- How to organise projects to deliver integrated flood risk management?

The following sub questions are derived from the main research question to guide the research:

Q1. What is integrated flood risk management and what are the related governance challenges?

Q2. How can governance arrangements support the delivery of integrated flood risk management projects?

Q3. How can projects contribute to a regime transition to integrated flood risk management?

1.3 Research approach

This research has studied a number of projects that have adopted an IFRM approach. These projects, used as case studies (Section 1.3.1), have provided the opportunity to obtain empirical evidence and an in-depth view of: the governance challenges (Q1); the functioning of governance arrangements (Q2) and the projects' contribution to a transition (Q3). The case studies have been conducted subsequently to each other. Theories and insights have been developed in the initial case studies and these have been validated and enriched using subsequent case studies. During the research the case studies have been examined using existing and emerging theories on: collaborative planning; social learning; IFRM; adaptive co-management; cross-sectoral collaborations; and transitions. This work has then been used to develop and enrich theoretical frameworks to deliver IFRM projects and for projects to contribute to a transition to IFRM. Figure 1.1 shows the research approach and distinguishes the existing theories (depicted in in white boxes) that have been used to examine the various case studies (depicted in dark grey boxes) and the results this research has generated in terms of the practical and scientific contributions (depicted in grey boxes). For each element of the research approach Figure 1.1 indicates the corresponding chapter where it is addressed (in a small black text box).

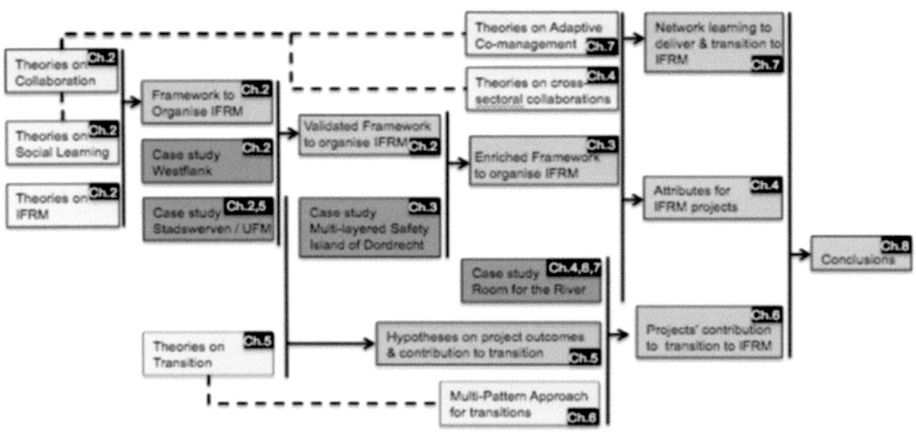

Figure 1.1. Research approach: theories; case studies and this research' contribution per Chapter

Table 1.1 further clarifies Figure 1.1 by presenting an overview of the Chapters of this thesis and the theories, case studies and practical and scientific contributions presented in each Chapter

Table 1.1. Chapters of this thesis and the theories, case studies and practical and scientific contributions presented in each Chapter

Chapter	Theories	Case studies (Section 1.3.1)	Practical and scientific contributions
2	Collaborative planning; social learning; IFRM	Westflank; Stadswerven	Validated Framework to organise IFRM projects
3	Framework to organise IFRM projects; IFRM	Multi layered safety Island of Dordrecht	Enriched framework to organise IFRM projects
4	Cross-sectoral collaborations; Framework to organise IFRM projects	Room for the River	Project attributes to foster and implement IFRM
5	Transitions	Stadswerven, Urban Flood Management	Classification of project outcomes and their contribution to a transition to IFRM
6	Transitions; Multi-Pattern Approach to describe transitions	Room for the River	Projects' contribution to a transition to IFRM
7	Adaptive co-management; network learning	Room for the River	Network learning to deliver IFRM projects and contribute to a transition

1.3.1 Case studies

The research presented in this thesis has examined four case studies in the Netherlands. They have been selected for several reasons. They have all adopted an IFRM approach, but they have each developed different options to reduce flood risk.

- Stadswerven: a flood proof urban development of a neighbourhood in the outer marches of Dordrecht.
- Westflank: creation of water storage capacity in a regional urban development in Haarlemmermeer.
- Island of Dordrecht: used the multi-layered-safety approach to combine flood protection, spatial planning and emergency management to increase the flood safety of the polder area protected by a ring dyke.
- Room for the River (RftR): delivered river widening measures to increase the river discharge capacity of the rivers Rhine, Meuse, Waal, IJssel and Lek.

Table 1.2. summarises the case studies from this research and the scope of the measures they comprise in relation to the physical system. It also presents the scope of traditional flood management to help understand the difference with IFRM that comprises a broader portfolio of measures. The images in Table 1.2 graphically represent the physical system in which IFRM can make interventions through flood risk reducing measures. It comprises (from right to left): an urban area with: assets (represented by a house in the image), inhabitants (represented by an icon of a man) and nature (represented by a tree); a rain cloud that represents the hazard for pluvial flooding; an urban drainage system underneath the urban area to reduce the probability of pluvial flooding; a dyke that protects the urban area from fluvial and coastal flooding; the unembanked area or outer marches with assets, inhabitants and nature; the riverbed; and finally another dyke on the other side of the river. The scope of the proposed measures and of the IFRM concepts in the case studies is indicated with a circle.

Table 1.2. The case studies from this research and the scope of the measures they comprise in relation to the physical system

Case Study	Scope
Traditional flood management: no case study included in this research on traditional flood management	
Stadswerven: flood proof urban development of a neighbourhood in the outer marches of Dordrecht	
Westflank: water storage capacity in regional urban development in Haarlemmermeer.	
Island of Dordrecht: multi-layered-safety (MLS) that combined flood protection, spatial planning and emergency management to increase the flood safety of the polder area protected by a ring dyke.	
Room for the River (RftR): delivered river widening measures to increase the river discharge capacity.	

The research presented in this thesis does not evaluate the benefits of IFRM and does not compare these to traditional flood management. Hence, no case study that has merely adopted traditional flood management has been included in this thesis. In fact, the measures (e.g. dykes) and methods (e.g. flood probability analysis) of traditional flood management form part of IFRM that comprises a larger portfolio of measures and methods. This research has circumvented the normative debate by merely using case studies that have used an IFRM approach. The relevance of the research is illustrated by many new projects that, despite a lack of evidence of benefits (Jeffrey and Gearey, 2006) and clear guidance (Section 1.1.3) adopt an IFRM approach (Butterworth et al., 2010).

All of the case studies were situated in the Netherlands. This facilitated the comparison of case study results for the development and implementation of various types of IFRM measures as the historic (flood history); cultural and socio-economic; and institutional contexts are the same. The lessons of this research are thus not necessarily transferrable to other countries and contexts and will require international validation. It has been beyond the scope of the research presented in this thesis to include case studies from other countries.

The case studies selected enabled the author of this thesis to take on different roles and perspectives, ranging from external observer to project coordinator, in relation to the empirical data from the various case studies (See Chapter 1.3.2).

The case studies were conducted subsequently to each other and thus advances of scientific literature and of insights by the researcher himself from initial case studies could be used for the research design and examination of subsequent case studies. A process framework to organize projects for collaboration and learning has been developed based on the Stadswerven and Westflank case studies. The framework has been applied and validated using the case studies of the Island of Dordrecht and of Room for the River. Based on the Stadswerven case study, insights have been gained on what outcomes the project generated that contributed to a regime transition. These insights have been examined using transition theories. The contribution of Room for the River to a transition to IFRM has been analysed to enrich these theories. Finally, the learning outcomes have been studied as part of the adaptive co-management process in Room for the River that has been instrumental in delivering integrated outputs and in contributing to a transition to IFRM.

1.3.2 Research context and role of researcher

The case studies have been part of 4 research projects in which the author of this thesis has worked together with practitioners, policy-makers and scientists. Table 1.3 describes the role of the author of this thesis in each case study. The projects Stadswerven and Westflank were supported by Learning and Action Alliances (LAAs) funded by the Dutch research programme Living with Water (LMW, 2010), Urban Flood Management Dordrecht (UFM, 2008; Van Herk, 2008) and Building with Water (BMW, 2008) respectively. The third project Multi-Level Safety for the Island of Dordrecht, was supported by an LAA funded by the Interreg 4b project Managing Adaptive REsponses to changing flood risk (MARE, 2011; Van Herk et al., 2011c). For the 2 research projects and LAAs related to Stadswerven and Dordrecht, the author was the project manager and LAA coordinator: steering research and demonstration work of researchers involved, practitioners and policy-makers; managing the interfaces between the various project activities; and with the context comprising other related research projects, policy processes and urban development projects. Also he was involved as external consultant to the project manager of the Westflank project and coordinator of the LAA Building with Water to advice them on the design of the collaborative planning process. Hence, for the first three case studies, the author of this thesis could operate in the role of an action researcher (See Section 1.3.3 for a critical reflection on his role of action researcher and on the scientific rigour regarding the methods he has applied in the research presented in this thesis). He designed and coordinated the project organisations and collaborative planning processes for these three case studies. He could thus test if the governance arrangements he implemented supported the delivery of IFRM and if the projects contributed to a transition.

Table 1.3. The role of the author of this thesis in each case study

Case Study	Role of author of this thesis in each case study
Stadswerven: flood proof urban development of a neighbourhood in the outer marches of Dordrecht	**Initiator** and **coordinator** of collaborative research project UFM and Learning & Action Alliance that supported Stadswerven. **Co-initiator** of flood-proof pilot project in Stadswerven. **Researcher** of (i) the LAA organisation and the collaborative process he designed and coordinated himself and (ii) the outcomes of Stadswerven and LAA activities and their contribution to a transition to IFRM.
Westflank: water storage capacity in regional urban development in Haarlemmermeer	**External consultant** to the developer on collaborative planning and to the collaborative research project to support project management. **Researcher** of the process (ex-durante and ex-post).
Island of Dordrecht: multi-layered-safety (MLS) that combined flood protection, spatial	**Initiator** and **coordinator** of a policy pilot for Delta Programme; and of a collaborative research and demonstration project MARE with Learning & Action

planning and emergency management to increase the flood safety of the polder area protected by a ring dyke.	Alliance that supported the Island of Dordrecht to increase flood safety. **Researcher** of the process he designed and coordinated himself. He has also been involved in related case studies in Dordrecht (e.g. Van Herk et al., 2012c; Koukoui et al., 2013) that are not part of this Thesis.
Room for the River (RftR): delivered river widening measures to increase the river discharge capacity.	**External researcher** and **observer** of various stakeholder events during 2 years of RftR

The frameworks that have been developed in this research have been validated and enriched by studying a fourth case study, Room for the River, for which the researcher was an observer rather than a participant as in the other cases. The RftR Programme funded this case study research to obtain an independent scientific evaluation. An independent scientific board has steered the content of the research and a user panel has advised on the relevance and potential implications of the research. The author was supervised by two professors and has conducted this research together with another researcher and PhD candidate, Mr. J. Rijke. Empirical data on RftR was gathered collectively, but analysis was done independently and from different theoretical perspectives. Mr. Rijke has investigated programme management and strategic planning to deliver adaptation of the physical system (e.g. Rijke et al., in press; under review), whereas this thesis focus on the organisation of the collaborative planning process in projects to deliver adaptation of both the physical and societal system. Together with other authors, Mr. Rijke and the author of this thesis have delivered to date 8 journal papers and 5 conference papers based on the RftR case study. Three of these journal papers have been included in this thesis (See Section 1.4 for a list of the papers that have been included and for the contribution of the author of this thesis to each of these papers) and 3 other papers have been included in the thesis of Mr. Rijke.

1.3.3 Overview of methods and critical reflection on action research

The case studies have used various methods that will be described per case study in each chapter. In all cases these comprised multiple methods to enable data triangulation. The methods have been summarized in Table 1.4 and the method 'action research' has been described in more detail below.

Table 1.4. A summary of the methods used in the case studies

Method	Stadswerven	Westflank	Island of Dordrecht	Room for the River
Action research	✔	✔	✔	X
Document analysis	✔	✔	✔	✔
Observation	✔	✔	✔	✔
Respondents of semi structured interviews and group interviews	24+36	24+36	17	55+13
Survey respondents	X	X	42	151
Respondents of validation interviews or workshops	52+35	52+35	6	220+150 approx.
Validation observation	X	X	X	✔

Action Research (AR) can be considered an important method that has been applied in the first 3 case studies, because the researcher was also a participant in the collaborative process and thus engaged in delivery as well as observation (Checkland and Holwell, 1998; Flood, 1998). The very essence of AR is encapsulated within its name: it represents a juxtaposition of action and research, or in other words, of practice and theory. Thus, as an approach to research, AR is committed to the production of new knowledge through the seeking of solutions or improvements to ¨real-life¨ practical problem situations (McKay and Marshall, 2001) by working collaboratively with other concerned and/or affected actors (Checkland, 1991; Hult and Lennung, 1980). In this research, the objective of applying this method has not only been to advance scientific knowledge (Whyte, 1989), but also to contribute to the delivery of the IFRM projects and to a transition to IFRM by e.g. influencing policy processes and developing new methods. The use of action research also generated new scientific insights as it gave access to more data; to knowledge and expertise of practitioners and policy makers (Burns, 1994); and a richer understanding and narrative on the relationship between the governance arrangements, collaborative processes and the outputs and outcomes of the case studies. Moreover, the underlying planning process for the case study of Island of Dordrecht could be designed using the researcher's framework; to be tested posteriorly. As stated by AR literature (e.g. Checkland, 1991; Baskerville and Wood-Harper, 1996), it is more than just another approach to problem solving, for the action researcher is working from within a conceptual framework. In fact, the role of action researcher stimulated the use and combination of multiple scientific theories, such as on collaborative planning, social learning, adaptive co-management and transition management in one framework, as these theories merge in practice (Section 8.4).

AR is not without its critics, as the researcher is not an impartial spectator on the research context (Chalmers, 1982), but participates and is thus biased. The author is aware of criticisms and limitations of AR and his presumed subjectivity and has attempted to overcome this by the elaborate application of multiple validation methods. Part of the research data have been obtained and analysed in parallel by other researchers, notably Mr. Jeroen Rijke, Mr. Chris Zevenbergen and Mr. Richard Ashley, for later comparison. Case study results have been validated by project participants and by external experts through interviews, workshops and surveys that were not involved in previous rounds of data collection and analysis that led to the results. The results have been tested on different case studies where the researcher had a different role in a different configuration of stakeholders. And the results have been compared with scientific literature.

Case study research is also criticised as it is difficult to draw generalizable conclusions. However, the case study approach can help the development of theory (George and Bennet, 2004; Brower et al., 2000), especially relevant here, as there is limited research available on the organisation of IFRM projects. The case study research followed the 'roadmap' of Eisenhardt (2003) and used recommendations of Flyvbjerg (2006), as theory development took place in parallel with the case study analysis. Also, the case study analysis used a triangulation of methods, following the principles of Yin (2003), so as to generalise and substantiate results (Webb et al. 1966). Triangulation is the combined use of multiple scientific methods to study the same phenomenon (Denzin, 1978). Jick (1979) recommends a combination of quantitative and qualitative methods for the validation process, which has been done as explained previously. This thesis can also be rightly criticised as the underlying research is not replicable. This is not only because of the role of the action researcher, but also because the projects studied are confined to a period in time and because the societal and physical systems in which they are delivered are changing or being changed during that period of time. In fact, the researcher has tried to contribute to the delivery of the projects and a transition to IFRM, with the ambition that future IFRM projects can be delivered more effectively and in a more supportive societal and governance context.

1.4 Structure of thesis, list of papers and contribution of the author

This thesis is comprised of eight chapters. Chapter 1 comprises this introduction. Chapters 2-4 analyse governance strategies for IFRM projects to deliver integrated outputs. Chapter 5 and 6 analyse how projects can contribute to a transition to IFRM. Chapter 7 combines delivery and transition through learning outcomes as part of adaptive co-management. Chapter 8 provides answers to the overarching research questions as presented in this chapter.

The body of this thesis (Chapters 2-7) consists of six papers that are under review, in press, or have already appeared in peer-reviewed journals. The intention of each paper is that it is sufficiently self-contained, so as to be understandable without recourse to the other papers. As a result, there is some overlap in content between the various papers, particularly regarding the introductions, the approaches and methods used and in some cases results are repeated. Table 1.5 lists the chapters and the related scientific papers. The scope of the chapters will be described in more detail below.

Table 1.5. The chapters of this thesis and the related papers

Chapter	Paper Reference
2	Van Herk, S., Zevenbergen, C., Rijke, J., Ashley, R., (2011) Learning and Action Alliances for the integration of flood risk management into urban planning: a new framework from empirical evidence from The Netherlands. *Environ. Sci. Policy*, 14, 543-554
3	Van Herk, S., Zevenbergen, C., Gersonius, B., Waals H., Kelder, E.T.G. (2013) Process design and management for integrated flood risk management: exploring the Multi Level Safety approach for Dordrecht, the Netherlands. Netherlands. *Journal of Water and Climate Change,* doi: 10.2166/wcc.2013.171
4	Van Herk, S., Rijke, J., Zevenbergen, C., Ashley, R., (under review) Attributes for integrated Flood Risk Management projects; case study Room for the River. Under review at: *International journal of River Basin Management*
5	Van Herk, S., Zevenbergen, C., Rijke, J., Ashley, R. (2011). Collaborative research to support transition towards integrating flood risk management in urban development, *Journal of Flood Risk Management*, Volume 4, Issue 4, December 2011, Pages: 306–317
6	Van Herk, S., Rijke, J., Zevenbergen, C., Ashley, R., (2013) Understanding the transition to integrated flood risk management in the Netherlands. *Journal of Environmental Innovations and Societal Transitions (EIST),* doi: 10.1016/j.eist.2013.11.001
7	Van Herk, S., Rijke, J., Zevenbergen, C., Ashley, R., Besseling, B. (in press) Adaptive co-management and network learning in the Room for the River programme. *Journal of Environmental Planning & Management.* doi: 10.1080/09640568.2013.873364

The contributions of the author of this thesis to all the 6 papers comprise: the research design; execution of the data collection and data analysis; interpretation of results; and the preparation and submission of the paper. Co-authors have contributed to these papers through the data collection (notably Mr. J. Rijke for the RftR case study and related papers) and through the critical revision of draft papers (notably Mr. C. Zevenbergen and Mr. R. Ashley). 15 other scientific papers have been produced with varying types of contributions of the author of this thesis during his PhD research, but have not been included in this thesis (See Section 9, References): Koukoui et al, 2013, Rijke et al., 2012a; 2012b; in press; under review; Zevenbergen et al. 2008a; 2008b; 2013a; 2013b; Dudley et al., 2013, Van Herk et al., 2013c; 2012a; 2012b; Gersonius et al., 2007; Waals et al., 2011.

Chapter 2 presents the case studies Stadswerven and Westflank. It presents a social learning framework, Learning and Action Alliance, to support collaborative planning and integrated flood risk management. The framework comprises 3 types of joint activities: system analysis; collaborative design; and governance. These supported demonstration projects through 3 threads by: establishing facts; creating images; and setting ambitions. This was done via 3 streams by: addressing problems; developing solutions; and influencing politics. The new framework has been demonstrated to provide an effective guide and new analytical tool to the organisation of a LAA and collaborative planning process for IFRM.

Chapter 3 presents the case study Island of Dordrecht where the new multi-layer-safety (MLS) approach has been applied in the context of the Delta Programme. MLS comprises 3 flood safety layers to reduce flood risk: flood protection; spatial planning; and emergency response. The definition of IFRM is further refined. And the developed framework for process design and management (from Chapter 2) has been used and evaluates whether or not the collaborative planning process led to an IFRM plan. The framework has been shown to be effective in the delivery of an IFRM plan; it has been enriched by defining the interfaces between and phasing of planning activities; and can be further improved to better guide implementation and governance activities.

Chapter 4 uses the definition of IFRM of Chapter 3 and the framework of Chapter 2 to analyse the Room for the River (RftR) case study. It devises attributes of effective investment projects in developing and implementing IFRM. Two types of attributes emerged: those that directly foster IFRM processes and outputs and those that enable the implementation of IFRM projects in practice. They can be instrumental in designing future research on IFRM and support the development of integrated plans as well as their implementation in practice through IFRM projects.

Chapter 5 goes back to the Stadswerven case study and the Urban Flood Management collaborative research project that supported the Stadswerven development project to study its contribution to a transition to IFRM. It demonstrates that the process of research, if collaborative, can contribute to the better integration of flood risk in urban planning. Collaborative research can support demonstration projects and their wider uptake, as well as policy development and subsequently a transition to integrated flood risk management. It also provides greater freedom to use different approaches, consider a broader range of problems and solutions, bring stakeholders together and facilitate capacity building. This paper calls for collaborative research that facilitates social learning and is impact focussed, preferably supporting innovative demonstration projects.

Chapter 6 studies the outcomes generated by the RftR programme that have an impact beyond the scope of the programme and that are sustained after the delivery of the programme. The Multi-Pattern Approach (MPA) has been applied for the first time to understand the dynamics of the transition to integrated flood risk management in the Netherlands. The detailed analysis of the outcomes of RftR provided in-depth evidence of how the transition occurred in practice. Lessons have been drawn to enrich the MPA that can help the further development and application of the MPA framework. A tentative application of MPA and extrapolation of lessons from RftR provides insights for policy makers and transition scientists to monitor the yet unknown outcomes that the new Delta Programme will have to generate to overcome new transition conditions and shaping a new chain of transition patterns.

Chapter 7 used the RftR case study to increase understanding on how learning takes place and can be stimulated within a programme. It shows that a programme is a versatile governance arrangement to embed a wide range of strategies for learning and adaptive co-management, whilst delivering environmental objectives.

Chapter 8 combines the results of this thesis to answer the research questions that, in Chapter 1, were derived from the general objective. The conclusions subsequently define IFRM and frame the governance challenges; and deduce lessons on how to organise for IFRM and attributes to foster and implement IFRM projects. It reflects on existing scientific theories and this thesis' contribution.

Chapter 2

Learning and Action Alliances for the integration of flood risk management into urban planning: a new framework from empirical evidence from the Netherlands

Chapter 2 Learning and Action Alliances for the integration of flood risk management into urban planning: a new framework from empirical evidence from the Netherlands

This chapter has been published as:

Van Herk, S., Zevenbergen, C., Rijke, J., Ashley, R., (2011) Learning and Action Alliances for the integration of flood risk management into urban planning: a new framework from empirical evidence from The Netherlands. *Environ. Sci. Policy*, 14, 543-554

Abstract

Urban development and regeneration present windows of opportunity to reduce flood vulnerability that are often not taken advantage of. Collaborative planning is needed to integrate planning and flood risk management and can be achieved by a social learning framework: Learning & Action Alliance (LAA). This paper presents a new framework on how to organise a LAA to support collaborative planning. The framework is verified based on empirical evidence from 2 case studies in the Netherlands where LAAs supported the adoption of an integrated approach to flood risk management and urban development. More than 60 interviewees reported that the LAA helped develop and applied relevant knowledge in 3 types of joint activities: system analysis; collaborative design; and governance. These supported demonstration projects through 3 threads by: establishing facts; creating images; and setting ambitions. This was done via 3 streams by: addressing problems; developing solutions; and influencing politics. The new framework has been demonstrated to provide an effective guide to the organisation of a LAA and provides a new analytical tool to assess the impact of LAAs. Other success factors for LAAs and the better integration of flood risk management into the planning process are considered.

2.1 Introduction

Urban floods are increasing worldwide and are likely to become even more damaging in future due to climate change (Munich Re, 2009). Increasingly it is recognised that the use of large infrastructure alone to combat this has the risk for technological lock-in and is likely to be less effective than integrated approaches to manage flood risk (Evans et al., 2004). Therefore a shift to an approach that comprises both structural and non-structural responses is going to be needed that also maximises multifunctional opportunities for land use. Amongst the non-structural responses, land use planning is considered as one of the more crucial components in managing flood risks (Wheater & Evans, 2009; White, 2010). An integrated approach to flood risk management set within urban planning processes is now seen as an effective way of minimising risk, although this has not always been recognised in practice and empirical guidance on its implementation is still lacking (e.g. DCLG, 2009; Carter et al., 2005). In certain countries, such as England, planning incorporating flood risk has become a mandatory requirement (since 2001) and takes a regional as well as a local perspective (DCLG, 2009). In the Netherlands the Dutch spatial planning procedure 'Room for the River' explicitly aims to increase flood safety combined with increased spatial quality of landscape, nature and culture (Ministry of Transport, Public Works and Water Management, 2000; Schut et al., 2010). In addition to planning for new developments, the large stock of existing buildings provides significant opportunities for retrofitting as part of normal redevelopments. Thus it may be possible to incrementally reduce flood vulnerability in European cities by taking advantage of current redevelopment opportunities. For this to happen, flood risk management needs to be integrated better into planning and urban retrofitting and development processes (ECTP, 2005).

However, there are many barriers to integrate flood risk in planning. Urban planning sets out to integrate different needs and requirements at a range of spatial and temporal scales. In this, flood risk management is not normally considered to be the most important of the various utility and service needs and opportunities. 'External integration' of wider priorities is part of a general ambition to make planning processes more inclusive in many developed countries (Spit & Zoete, 2006). Flood risk management is a long term goal compared with many other planning considerations. To address this Hajer et al. (2006) suggest a transition to 'horizontally, interactive' planning, as opposed to 'vertically institutional' in order to foster flexible and dynamic planning regimes. These should be flexible and dynamic enough to address contemporary, complex challenges - such as flood risk - combining spatial quality with democratic legitimacy. The multi-objective decision making used in urban planning is complex and needs decision makers who are capable of planning urban areas that can accommodate uncertain futures. Actions are needed in the short term with short term benefits to bring about the required changes to deliver the longer term plans to make urban areas flood resilient (Geldof, 2007; Hamin & Gurran, 2009; White, 2010). Despite this, the complexity of planning processes means that no single stakeholder group has final or absolute control over urban or spatial developments (Sellers, 2002). This multi-actor setting further complicates the way in which flood risk can be adequately addressed in planning processes (OECD, 2010). Another set of barriers includes the increasing uncertainty about changing flood risk (Milly et al., 2008) and a lack of understanding or shared perception of the effectiveness of non-standard response measures (Adger et al., 2005). This lack of understanding also contributes to a technical lock-in to structural solutions, such as defence measures (Walker 2000). The barriers point towards 'wicked-problems'; being "problems that have multiple and conflicting criteria for defining solutions, solutions that create problems for others, and no rules for determining when problems can be said to be solved" (Rittel and Webber, 1973) or persistent problems that are ill-structured, involve many stakeholders, are surrounded by structural uncertainties, and are hard to manage (Rotmans, 2005).

Integrated, collaborative planning should overcome these barriers and involves facilitating complex, decision making. Governance and network theories (Hanf and Scharpf 1978; Kickert et al.,1997; Marsh and Rhodes, 1992; Rhodes 1997; Scharpf 1997) indicate that stakeholders are becoming more actively involved in decision making to develop a joint definition of the problems and potential responses and for the sharing of interests, aims, and ambitions; and also to learn together. Interactive decision making is expected to result in richer policy proposals that can be implemented more efficiently and thus raise the democratic legitimacy of the decisions (Edelenbos & Klijn, 2006; Farrelly et al. 2009). More persistent barriers, such as: institutional structures, cultures and approaches in planning and flood risk management, require a transition or regime change (e.g. Pahl-Wostl, 2007; van der Brugge and Rotmans, 2007; Wong and Brown, 2009). Also these transitions require increased and wide stakeholder involvement to help reach consensual decisions (Rhodes, 1997; Pahl-Wostl et al., 2007; Ashley et al., 2012). The need for integration of flood risk management into the way in which cities are planned, requires new skills and competencies in all of the stakeholders concerned and they need to be better supported (Evans et al, 2004). Pahl-Wostl et al. (2007) recommend "social learning" processes as a means of developing and sustaining the capacity of different authorities, experts, interest groups, and the general public to manage their water systems effectively. Social learning includes both the capacity building of individuals and organisations, as well as the creation of relational qualities and social capital (Pahl-Wostl et al., 2007; Bouwen and Tallieu, 2004). Social learning also helps to build on experience to cope with uncertainty and change, which is especially relevant for integrated flood risk management and urban planning (Folke, 2006). In short, social learning is considered an alternative, complementary policy instrument in water governance (Blackmore et al. (2007) and is gaining recognition as a potential governance or coordination mechanism (Ison & Watson, 2007). Hence many authors stress the need for multi-party collaboration and propose some sort of social learning framework (White, 2008; Boelens, 2006; Pahl-Wostl et al., 2007; Farrelly et al., 2009, etc.).

The question arises as to how best to organise a social learning framework to support collaborative planning. Few authors provide guidance on how to organise a social learning framework and these are not focussed on supporting collaborative planning or flood risk management. Lipnack and Stamps (1997) propose key principles of networked organisations: unifying purpose; independent members; voluntary links; multiple leaders and integrated levels. Others (Franke, 1999; Miles and Snow, 1986), propose the duties of the net-broker or coordinator of such networks: initiation and preparation of the network; maintaining and improving the network collaboration; promoting the partnership concept; monitoring and continuously improving network performance; responding to opportunities. Senge (1990) presents learning objectives: building a shared vision; personal mastery; surfacing and testing mental models; team learning and systems thinking. Finally, Daniell et al. (2010) provides recommendations on the 'co-engineering' and negotiation process to set up the organisation to support a participatory water management process, but not on the eventual design of the organisation itself.

This paper describes and evaluates a new framework to organise social learning to support collaborative planning, drawing on theories on social learning, development planning (Van Buuren, 2006) and decision making (Teisman, 2000; Kingdon 1984). There are a number of potentially relevant case studies in which stakeholders and experts have been part of a new process of co-production and evolution aimed at the development and implementation of a shared, integrated and adaptive approach to manage flooding within the planning and development process. However, these have so far been poorly monitored and not effectively evaluated and documented (Pahl-Wostl, 2007). This paper evaluates two case studies. These case studies demonstrate how flood risk management has been supported by a social learning framework so as to be better incorporated into urban development planning.

2.2 Learning & Action Alliance and knowledge

Social learning mechanisms or the frameworks of multi-party collaboration necessary to deliver social learning have been variously named: communities of practice (Wenger, 2000); learning alliance (Verhagen et al., 2008; Batchelor & Butterworth, 2008); learning & action alliance (Ashley et al., 2012; Newman et al, 2011); socially-embedded institutions (Cleaver, 2002); learning platforms or arenas (Farrelly et al., 2009); learning networks for sustainable development (De Kraker et al., 2010; Von Malmborg, 2007; Manring et al., 2003); learning organisation (Senge 1990) and networked organisations (Lipnack and Stamps, 1997). Often without providing an explicit or unambiguous definition. These frameworks all have in common a multi-stakeholder and learning aspect. The differences are found mostly in their operational aims or in the transition stage they contribute to. Daniell et al. (2010) distinguish between a management-driven participatory process with instrumental goals and research-driven processes focusing more heavily on social learning. Both aims are important to integrate flood risk management and urban planning and thus in this paper we will use the term Learning & Action Alliance (LAA) of Newman et al. (2011). They use the definition of a Learning Alliance by Batchelor & Butterworth (2008): "a group of individuals or organisations with a shared interest in innovation and the scaling-up of innovation, in a topic of mutual interest", and add the word *Action* to highlight both its learning and also its delivery aspects. In the context of this research *Action* refers to the integration of flood risk management in urban development planning projects. Nonetheless an LAA can be used for different innovations in different sectors. Van Herk et al. (2011b) pose the hypothesis that different social learning frameworks are necessary at different stages of a transition and that collaborative research is appropriate during the early stages of transition. The integration of flood risk management and urban planning is in a early phase of its transition from 'fighting against water' to 'living with water' (Rijke et al, 2008; Newman et al, 2011) and needs innovative demonstration projects and the creation of networks to influence policy processes and change the regime (Van Herk et al., 2011b). However, also in later transition phases, continuous social learning is still necessary to cope with uncertainty and change (Folke, 2006), whilst participatory planning will always be needed for urban planning and flood risk management as policies are not set at a certain moment by a certain actor (Healey, 1998; De Bruijn & Ten Heuvelhof, 1999). A LAA can serve all stages of this transition.

The main output of LAAs is knowledge (Wenger, 2000). It is when this emergent, contextualised knowledge is coupled to carefully designed social interactions, that bring people into new relationships with each other and the resources at stake, that it is possible to envisage practice-driven policy processes, informed by a process of multi-stakeholder knowledge generation (Jiggins et al., 2007, p533). Knowledge has many forms, as described by Bläckler (1995): ¨knowledge is multi-faceted and complex, being both situated and abstract, implicit and explicit, distributed and individual, physical and mental, developing and static, verbal and encoded.¨ According to Van Buuren (2006) there are three categories of knowledge: (i) explicit, factual and impersonal; (ii) socially construed, normatively loaded reality definitions and images (Schön & Rein, 1994, Fischer, 1990); and (iii) experience-based competencies and skills (Schön, 1983; Cook & Brown, 1999). Tacit knowledge especially, covering the second and third categories, can contribute to the innovation (Nonaka & Taceuchi, 1995) that is necessary for demonstration projects and the transition towards integrated flood risk management. The knowledge development in LAAs comprises all these types of knowledge. By means of co-generating knowledge and stimulating its application, a LAA can reach its aims to support integrated, collaborative planning, facilitate social learning and thus support a wider transition to effectively include flood risk management in development planning processes. Complex decision making –as integrated planning- requires applicable knowledge (Lindblom & Cohen, 1979). Knowledge needs to feed into planning and land-use decisions and is the basis of any flood resilient city (White, 2008). Thus, the organisation of a LAA should enable the development, exchange and application of knowledge.

2.3 Organising an LAA to support collaborative planning

Van Buuren & Nooteboom (2009) present 3 ways in which knowledge development can contribute to the quality of decision making in flood risk management and urban development planning. Knowledge development focussed on: (1) the quality of the ultimate policy choice (usefulness, applicability); (2) the procedural quality of the planning process (transparency, timeliness) and (3) the quality of stakeholder participation in the planning process (openness, equity, dialogue). These 3 *logics*, as they call them, can be complementary, seem often sequential and can be deliberately chosen, but often are not. Along similar lines Van Buuren (2006) studied in earlier work the role of knowledge in decision-making for spatial development processes and presented 3 interrelated *threads*:

1. to *establish facts*. This thread generates knowledge that: is coherent and not contradictory, has a proven quality and serves to reduce uncertainty and has

been established without unacceptable influence from the wishes and opinions of the parties involved;
2. to *create images*. This thread supports *frame reflection* in which parties identify their view of reality and discuss it, look for images or meanings that they share, and create renewed and more creative images as a result of the interaction;
3. to *set ambitions*. This thread supports the negotiations on aspirations of the parties towards implementation.

Thus, a LAA is to be organised to support collaborative planning via these three threads. A LAA should support the *fact* (1) thread by generating and exchanging factual knowledge, and the *image* (2) and *ambitions* (3) threads by bringing stakeholders together voluntarily and in a way that they can freely discuss interests and views. In this way a LAA can thus support interactive decision making. A LAA also supports different types of (social) learning: *single-loop* learning or capacity building through the fact thread; *double-loop learning* to change beliefs, norms and objectives through the image and ambition thread; and *deutero-learning*, learning the ability to learn simply by engaging in the learning process (based on: Tuinstra et al., 2008). In the case studies presented in this paper, the contribution of the LAAs to decision making via each *thread* has been analysed. The contribution of the LAAs to the various types of learning was not explicitly considered in this study.

How do these facts, images, ambitions actually end up in planning decisions? We have described how threads support decision making, but not how the knowledge is actually applied and captured in policy and business proposals. To stimulate the actual uptake of knowledge from a LAA requires an understanding of decision making processes in urban development planning. Ratcliffe et al. (2004) stress the importance of understanding the dynamics of the procedural milestones and political and corporate decision making processes in urban planning. Decision making processes in urban development seem unstructured as policies are not set at a certain moment by a certain actor (Healey, 1998; De Bruijn & Ten Heuvelhof, 1999). Kingdon (1984) presented the *stream model* to help understand such complex decision making processes. He defines decision making as the connection between three concurrent streams of problems, policies / solutions and politics or participants (Teisman, 2000). These streams are independent, each with their own dynamics. Linkages between the 3 streams occur if there is favourable momentum; a so-called 'policy window' (Kingdon, 1984). Following the stream model, an LAA is to be organised to:

(i) Analyse and address problems;
(ii) Develop and propose solutions; and

(iii) Influence politics by seeking political commitment or bringing participants together.

Thus, an LAA can push and pull the 'streams' of problems, solutions and politics to take advantage of, or even provoke, a policy window for decision making. Pahl Wostl et al. (2007) posed similar hypotheses related to problems; claiming that the processes of framing and reframing 'a problem domain' are essential elements of the social dynamics of the group during processes of negotiation of meaning. Farrelly and Brown (2011) highlight the importance of multi-actor collaboration and the role of scientific partners to develop new solutions to be applied experiments or demonstration projects in urban water management. Bringing participants together is an inherent characteristic of an LAA. The case studies here seek empirical evidence of the contribution of LAAs on the *streams* in the decision making process.

2.4 Framework for organising an LAA

In short, an LAA is to be organised to generate and apply knowledge to contribute to collaborative planning via the three threads: establish facts; create images and set ambitions, and the three streams in the decision-making process: address problems; propose solutions; and bring participants together. Additionally in this paper we propose a framework to organise a LAA around 3 groups of activities aiming to contribute to the threads and streams: 1. system analysis; 2. collaborative design; 3. governance. Professionals involved in collaborative planning or related social learning processes might implicitly work already according to these notions. For example, based on three existing approaches in the Netherlands, Van der Ven et al. (2006) recommend a mix of optimization, design and negotiation activities for urban water management planning. The framework we present here, however, logically structures LAA activities, their interfaces and relates them to objectives to support collaborative planning via threads and streams. Figure 2.1 illustrates the 3 LAA activities. The 3 activities (which actually comprise matrices) are presented in a loop and are interconnected by arrows as the activities are mutually supportive and run in parallel. Each activity is represented as a matrix that shows how the activity delivers in regard to the threads and streams (Sections 3) in 5 cells.

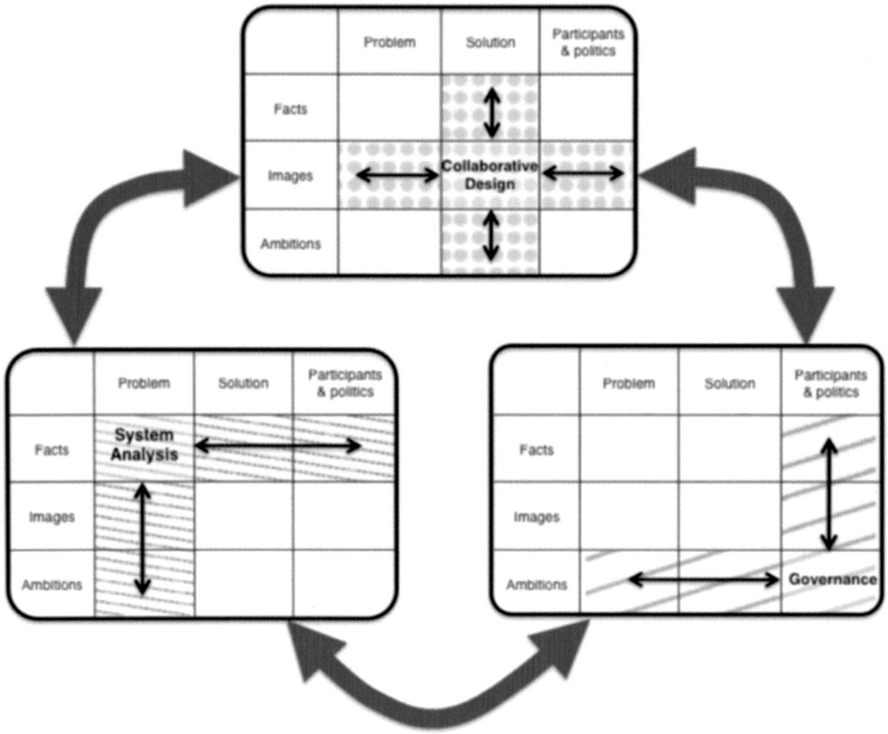

Figure 2.1. 3 interactive LAA activities contributing to collaborative planning via 3 threads (facts, images, ambitions) and 3 streams (problem, solution, participants & politics)

The 3 activities can comprise different work tasks that need to be run in parallel supporting each other. The interactions can be deduced from Figure 2.1 by overlaying the matrices of each activity, with overlaps occurring for four cells in each activity. System analysis influences design objectives and the range of design variables as stakeholders discuss the underlying objectives and problems. System analysis both involves participants in establishing facts and problems can be discussed. These can be prioritised and addressed amongst decision makers based on results from system analysis. Similarly, collaborative design involves participants in developing solutions that, in turn, feed into discussions on ambitions. The interests and ambitions of participants should guide design work, and solutions should be discussed. Successful interface management between collaborative design and governance can be assessed from *actor contentment* and *enrichment* (Edelenbos & Klijn, 2006); i.e. the degree to which the outcome of interactive processes is regarded as positive by actors, and the variety of ideas and the influence of these ideas on the outcomes such as decisions and plans. In this section the activities and their contribution to collaborative planning via threads and streams are described in more detailed.

2.4.1 System Analysis

System analysis mainly aims to ***establish facts*** and to analyse, define or reframe and address ***problems***. The relevant facts or information requirements for flood risk management include for example: definition of the areas at risk from flooding, green infrastructure mapping, ground composition, flow paths, nature and vulnerability of urban fabric, critical infrastructure (White, 2008). System analysis should also study the various layers of a spatial system: substratum, infrastructure networks and occupation patterns (Priemus, 2007) at various scales such as catchment, urban area and building level (Zevenbergen et al., 2008). It should involve participants in establishing facts jointly as all stakeholders possess knowledge, including factual knowledge from previous studies, relevant to frame problems and solutions, and to provide knowledge to decision making (Renn, 1995; Ehrmann & Stinson 1999; Petts & Brooks 2006). Moreover, decision makers have to balance multiple objectives or problems: e.g. flood safety has to be combined, or can compete with housing and many other objectives. All these objectives engage many and diverse stakeholders; each with their own interests. LAAs aim to create and exploit the freedom to consider a broad range of problems, stakeholder interests and solutions. The LAA members should share their experiences and knowledge in free-flowing, creative ways that foster new approaches to problems (Wenger, 2000). As a consequence there is more room for creativity and innovation because the participants are less likely to start to negotiate from entrenched positions (Pahl-Wostl et al., 2007).

2.4.2 Collaborative Design

Collaborative design mainly aims to ***create images*** and to develop, analyse, discuss and propose ***solutions*** and strategies. Collaborative design typically results in urban masterplan designs that can be developed in workshops with multiple stakeholders considering multiple objectives. In urban design, objectives can compete or may need to be combined. Therefore design is an activity recommended for frame reflection and to exchange perspectives (create images). Design work is a task oriented action with relational qualities of reciprocity and reflexivity recommended for social learning (Bouwen & Tallieu, 2004). Collaborative design includes the development of different alternatives instead of one single preferred design. Alternatives can show a larger variety of solutions to address multiple problems and allow participants to discuss these freely. Rapid prototyping can help to stimulate progress and help with the interaction with the parallel activities of system analysis and governance. LAAs are expected to be more useful when demonstration projects are in an early development stage when there is more flexibility in the design. In the evolving process the options available for decision makers will gradually decrease, whilst early design decisions generally constrain later change and have a high impact on the overall project investment cost (Van Herk et al., 2006). Flood risk is most effectively addressed in the early stages of urban development and in an integrated way (Carter et al., 2005, White, 2008).

2.4.3 Governance

The governance activity involves ***participants*** in discussing and defining their ***ambitions*** to create socially construed knowledge and highlight the ambitions and roles of the stakeholders regarding the development project and policy in general. The problems and solutions as defined in the system analysis and collaborative design are here discussed by the stakeholders. Complex decision making involves stakeholders with different views, stakes and values interpreting and valuing information differently; mobilizing different knowledge to underline viewpoints (Koppenjan & Klijn, 2004). The process of generating knowledge collectively in an LAA and discussing it in this activity can help address differences and avoid conflict and also preserve scientific validity (Lindblom & Cohen, 1979; De Bruijn & Ten Heuvelhof, 1999). 'Governance' also provides the vehicle to consider the existing institutional and policy framework and the roles and tasks of each stakeholder. If deemed appropriate, policy processes can be initiated and supported from the LAA for a broader transition in spatial planning or water management beyond the development project (Van Herk et al., 2009). Such social learning can support a change in the governance structure that may be required for general environmental improvement (Pahl-Wostl et al., 2007).

2.5 Methodology

The investigation used two case studies in the Netherlands to study the dynamics of the Learning & Action Alliances that were established in each. This was to develop a common understanding of the function and value of the LAA in ensuring that flood risk management was effectively included in the development planning processes. The case studies were selected to encompass different types of development and alternative sources of flood risk and for which the dynamics of the LAA in each could be expected to be similar: (1) De Stadswerven, an urban redevelopment project in Dordrecht, which has increasing fluvial flood risk from surrounding rivers; and (2) Westflank, an urban development project in Haarlemmermeer, with a need for increased freshwater storage and with increasing pluvial flood risk from the urban drainage system. The two projects were supported by LAAs funded by the Dutch research programme Living with Water (LMW, 2010), Urban Flood Management Dordrecht (UFM, 2008) and Building with Water (BMW, 2008) respectively. The studies of the collaborative planning processes and LAAs followed the 'roadmap' of Eisenhardt (1989) and used recommendations of Flyvbjerg (2006), as theory development took place in parallel with the case study analysis (section 5). The theory on the contribution of the LAAs to collaborative planning (section 3) was verified based on the first De Stadswerven case study. Then the relation was made with LAA organisation (section 4) and validated with the Westflank case study. The case study analysis involved a triangulation of multiple methods, following the principles of Yin (2003). The data comprised for each case study: a) 2 workshops with stakeholders that participated in the urban planning process and Learning & Action Alliance (n=36); b) semi-structured individual and group interviews (n=24). a) and b) focussed on the activities and results of the LAAs and the LAA contribution to collaborative planning and decision making. Questionnaires were filled in, then the researchers classified the answers in activities, threads and streams (section 4), after which the results were discussed individually and plenary. c) Observation as the researchers were also participants in the planning process and LAA and thus engaged in delivery as well as observation (action research: Checkland & Holwell, 1998; Flood 1998). d) Documentation content analysis of relevant policy, Stadswerven and Westflank project documentation, minutes of LAA meetings, media and scientific literature. e) A validation test was organised through a large workshop with external experts (n=52) and stakeholders from the LAAs studied and planning processes (n=35) to draw out contradictory and/or alternative explanations of the development and uptake of knowledge out of the LAA in the planning process. During this workshop also the relevance of the theory and case experiences were discussed.

2.6 Case Studies

2.6.1 Flood resilient waterfront development at risk from the river

The De Stadswerven case study was a redevelopment project in a former shipyard area, located on the edge of the historical city centre of Dordrecht. It was proposed to develop 1600 residential buildings with an associated range of commercial, cultural and public facilities. The 30 hectare development area was located in the outer flood plain which is not protected by the main ring dike for the City, being adjacent to the river confluence for three rivers: the Beneden-Merwede; Noord and Oude Maas; one of the most intensively used waterway areas in the world. Fluvial flood risk was initially not considered in the plans, but at the outset the requirement was stipulated to raise the ground level of all new developments in the area in accordance with the Dutch Water Act (2009) to ensure flood safety. Under influence of a LAA, the approach radically altered from one of 'flood prevention' to that of 'flood risk management' in common with initiatives elsewhere.

The development process started in 2001 when the municipality acquired land and developed an urban masterplan for De Stadswerven that was eventually approved by the city council in April 2005. Following the concept of flood defence, the ground level was to be raised up to 4m above mean sea level to meet regulations. In the following period the delivery of the De Stadswerven redevelopment was delayed and the approved masterplan was subject to further review due to a perceived lack of ambition and estimated cost overruns. This created tension between the private development consortium and the municipality and also political tension between the city council and government. Two local governors were questioned by the local parliament and the municipality faced a lawsuit against the private developers. The city government decided to suspend work on the project in September 2007. During the suspension of work on the project, there was a change of ownership in the private development consortium, now led by a new private developer. Meanwhile, the research project Urban Flood Management (UFM) Dordrecht (2005-2008) was created with a LAA comprising public, private and research partners. The De Stadswerven redevelopment was used as a case study for knowledge development for the LAA UFM, without initially directly engaging in the project; i.e. only as external observers. Nevertheless, after the project planning was temporarily suspended in 2007, the UFM LAA then had to directly support the delivery of the project. The UFM LAA challenged the requirement to raise the ground level of the entire area based on flood risk analysis suggesting a low vulnerability. Upon resumption, the masterplan was redefined which provided an opportunity, a 'policy window', to use an integrated flood risk management approach proposed by the UFM LAA. Flood risk was used as a design variable for the new masterplan. It included lower lying, attractive water-rich, yet flood proof areas as well as raised areas and infrastructure that can serve as safe havens and escape routes for evacuation. As a first step towards its implementation in practice, the UFM LAA proposed and actively participated in a feasibility study for a pilot FRM project to provide 100 flood proof dwellings in the De Stadswerven redevelopment.

2.6.2 LAA Urban Flood Management's (UFM) contribution to collaborative planning

The contribution of the LAA UFM to the decision making process for De Stadswerven is described in Table 2.1 using the framework as presented in section 4. For each activity type the impact is presented via threads (van Buuren, 2006) and streams (Kingdon, 1984). The activities are described in more detail on the UFM website (UFM, 2008) and by Gersonius et al. (2007).

Table 2.1. LAA UFM activities and contribution to collaborative planning of De Stadswerven via 3 threads and 3 streams

Activities: system analysis; collaborative design; governance.	Contribution on **threads**: facts; images; ambitions.	Contribution on **streams**: problems; solutions; participants / politics.
System Analysis consisting of: flood mapping (water levels & flow velocity); ground level analysis, vulnerability analysis of buildings, infrastructure; public space and critical assets; climate change scenario analysis. Moreover risk perceptions of inhabitants were studied, as well as the insurability of assets in De Stadswerven against flood risk.	The obtained **facts** on flood risk, showed that the redevelopment area was not particularly vulnerable to flooding compared with the rest of the city as it was situated on higher ground. This information was effectively communicated via visualizations in public presentations and publicatons that promoted public debate on flood safety in areas outside the main dikes, that were previously thought of as unsafe as they are unprotected by dikes. (**images & ambitions**)	The **problem** was reframed in discussion with stakeholders (**participants**) as was found that: flood risk is limited, it could be insurable and communication to inhabitants requires attention. These insights stimulated an investigation via the LAA for alternative design **solutions** that did not rely on elevating ground levels.
Collaborative design work on appealing designs of masterplans with attractive, water rich areas with marketable developments. Options for robust and adaptable spatial strategies for future uncertainties were investigated. Also a communication strategy was devised. Finally, design work for the assets in a pilot project of 100 flood proof dwellings.	Stakeholders got new views of reality (**images**) as the UFM LAA showed with her designs that flood risk is manageable through urban planning, flood proofing design and emergency response. The areas outside the dikes were presented as *safe havens*, not high risk areas. Furthermore, the design concepts provided options that potentially increased the spatial quality of the redevelopment whilst also reducing flood risk and addressing the project constraints (**ambitions**). The urban planners in Dordrecht	The proposed measures were integrated **solutions** that simultaneously addressed the interests of multiple stakeholders. The designs were considered inspiring, yet were not considered threatening to their own plans by the overall De Stadswerven project management team as long as they lacked detail. The new masterplan for the redevelopment was made by the same urban planners (**participants**) that participated in the LAA and ultimately incorporated several flood risk reducing measures designed in the UFM project. It included 'flood

	believed that these measures not only addressed flood risk, but created a more attractive living environment as the relation with the water becomes more apparent and the area has a variety of different ground levels.	free routes' and water-rich areas where the tidal influence can be lived with by inhabitants and thus account appropriately for climate change.
Governance. The LAA discussed the results from system analysis and collaborative design. UFM brought together a broader consortium of stakeholders than had originally been used for project delivery of De Stadswerven. This group launched the idea for a pilot project with 100 flood proof dwellings. An inventory was made of existing policies related to FRM and urban development in areas outside the dikes. Responsibilities and liabilities of different stakeholders were discussed and policy recommendations were drafted.	The pilot project with 100 flood proof dwellings made discussions on **ambitions** explicit as the concepts were brought to practice. The project was presented and seen externally (**images**) as an example of innovative planned building, which is both attractive and cost-effective in areas outside protective dikes. The LAA explicitly mobilised external political and financial support for implementation of the solutions proposed by UFM, providing opportunities for the De Stadswerven project managers (**ambitions**). E.g. the LAA UFM lobbied for financial and procedural support from the national DeltaTechnology Committee as the new masterplan might require additional investment and increase process and legal complexity.	Decision makers, including the city alderman and regional and national governments representatives formed a steering committee of the LAA that could support project delivery (**participants / politics**). The LAA UFM initiated and supported national and regional policy change for governance of developments in areas outside the dikes and, in general, for integrated flood risk management. E.g. the new policy of the Province of South Holland used UFM pilot as an exemplar. The integrated flood risk management approach attracted interest and support from the communication media and thence attention at Dutch Government level. Prior to this, the project had received mainly negative media attention due to the political tensions and project delays. Later, the UFM findings were adopted by the national Delta Commission and the Dutch Crown Prince visited Dordrecht.

2.6.3 Water storage in high density urban areas in the Netherlands

The second case illustrates the initiation and design phases of a development in the Westflank Haarlemmermeer area of the Netherlands. The area is situated in the Randstad and metropolis region of Amsterdam and is expected to provide some 20,000 new dwellings between 2010 and 2030. The planned development is to take into account a number of spatial constraints: spatial and environmental policies for Amsterdam Schiphol airport; spatial policy for the Greenport Bollenstreek area and water policy to increase local water detention and storage. The development comprises 3036 hectares, with 10,000 new dwellings; 1M m^3 peak flood retention capacity; 2M m^3 seasonal flood water storage and some 900 hectares of recreational green space. The LAA provided a new approach to effectively deliver on the water storage objectives that is now setting an example for profitable, attractive developments. This, in turn, is changing policy in The Netherlands in relation to such developments.

From 2004 a spatial plan was developed for Haarlemmermeer Bollenstreek, adopted by regional parliaments and presented to the national ministry of Housing, Spatial planning and the Environment in May 2006. Within a year of the start of the spatial planning process for the proposed development, a broad consortium of public, private and research partners created an LAA as part of the research project 'Building with Water' (BwW). BwW (2004-2008) aimed to learn about flood resilient and climate proof urban development in polder areas - behind the dike rings - and had the explicit ambition from the outset to contribute to the overall planning of the Westflank development. The LAA BwW identified the Westflank as a possible location for water storage and housing, for which a spatial exploration was executed with support from BwW. This included conceptual designs of high density urban areas with water storage, supported by a technical feasibility study, a market study and a cost benefit analysis. In November 2007, after lobbying of the LAA BwW, the project was included in the national policy programme Urgency Programme Randstad, thus providing access to additional financial legal and political support. In the following 2 years a development strategy was defined with water areas and multi-functional use of space as guiding design principles. As a first step towards its implementation in practice, BwW stakeholders agreed on the intention to deliver a pilot project 'waterliving' of 500 dwellings.

2.6.4 LAA Building with Water's (BwW) contribution to collaborative planning

The contribution of the LAA BwW to the decision making process for Westflank is described in Table 2.2 using the framework as presented in section 4. For each activity type the impact is presented via threads (van Buuren, 2006) and streams (Kingdon, 1984). The activities are described in more detail on the BwW website (BMW, 2008).

Table 2.2. LAA BwW activities and contribution to collaborative planning of Westflank via 3 threads and 3 streams

Activities: system analysis; collaborative design; governance.	Contribution on **threads**: facts; images; ambitions.	Contribution on **streams**: problems; solutions; participants / politics.
System Analysis consisting of: an analysis of the (ground) water-soil system (incl salt water intrusion) to verify the need and possibilities for water storage: flood mapping, inventory of civil works and infrastructure, landscape. LAA BwW conducted a real estate market study for 'water living' on potential buyers, needs, products, and prices; and a cost benefit analysis for the exploitation of water living.	The aggregated spatial aspirations from all of the original objectives exceeded the available space as illustrated by the BwW project, with a map showing the spatial demands. (**facts**). Based on that BwW redefined the required water storage objectives from a plan-based, footprint approach in hectares, to a volumetric storage approach in cubic meters; thus opening the way for a consideration of locations for water storage with different water depths. (**image / ambitions**) The market study found that new building typologies in the area were also competitive in terms of cost and market demands (**facts**).	LAA BwW showed the need and possibilities for water storage **(problems)** which stimulated the designers to investigate the possibilities of combining functions. (**solutions**). The **problem** definition was broadened with insight on pluvial flood risk as well as salt water intrusion. The marketability and cost-effectiveness of 'water living' raised the interest of private developers and contractors (**participants / politics**).
Collaborative design. LAA BwW developed conceptual and detailed designs for high(er) density urban environments with water storage applicable in various areas. LAA BwW developed a housing-water-index to show the possible housing density when combined with water storage that was used as input for design work.	The LAA BwW showed the potential of multi-functional use of space (**image**), particularly of high density urban development combined with water storage. The developed designs by BwW were considered appealing and functional by Westflank Haarlemmermeer stakeholders (**image, ambitions**).	BwW provided **solutions** showing the potential of multi-functional use of space with a newly developed housing-water-index. Based on that concept the LAA supported the zoning and location search for water storage. The Westflank was proposed as an area where high density urban development and thus with high expected revenues could be combined with water storage designed for fluctuating water levels. The latest masterplan designs incorporate many examples; e.g. sub-areas with a variety of urban 'amphibious' structures such as

		flood resilient homes, infrastructure and public green spaces. (**solution**)
Governance. LAA BwW brought together a broad consortium of stakeholders, especially enriched by private stakeholders. The LAA BwW provided the means to lobby widely for broad public, political and financial support. E.g. LAA BwW connected with senior national policy makers and politicians and successfully lobbied for inclusion of Westflank in Urgency Programme Randstad. The LAA BwW proposed and supported a feasibility study for 'waterliving' pilot of 500 dwellings.	The LAA BwW considered a broad range of sectoral and stakeholder **ambitions**; especially interest of land owners, developers and contractors were brought into a public driven process. Urgency Programme Randstad gave a boost to the project and paved the way for an intention agreement for a pilot project 'water living' in November 2007 signed by all major stakeholders. (**ambitions**).	The LAA introduced new stakeholders and interests to the process (**participants/politics**). This supported **problem** reframing, e.g. including private interests. Support for new **solutions** was forged towards implementation through 'waterliving' pilot project and further backed by the Urgency Programme Randstad, providing access to additional financial, legal and political support for the urban development (**politics**).

2.7 Discussion and Analysis

This research has revealed that LAAs can support collaborative planning. The case studies in Stadswerven and Westflank provided relevant empirical evidence of social frameworks that supported the integration of flood risk management into urban planning processes. The two LAAs have had a decisive influence on the urban masterplans and related policy proposals that have been adopted. The LAA UFM proposed design concepts with varying ground levels and a new flood proof building typology. BwW proposed the combination of housing and green areas with water storage to exploit fluctuating water levels. The design concepts embraced multifunctional use of space through which flood risk management was connected with other objectives and interests. For both case studies, respondents highlighted a shift in perspective where flood risk management was seen as a solution instead of a problem.

The LAA brought together a broad range of stakeholders each with their interest and expertise that steered and enabled the reframing of the problem and the development of these innovative solutions. Participants confirmed that the prospect of a learning experience attracted them and more participants were motivated to engage in a more open dialogue on ambitions than in the traditional urban development processes. In interviews with the stakeholders, spatial design was seen to stimulate fruitful discussion as all objectives needed to be merged into a single holistic urban design. One respondent insightfully called the proposed designs "lucky dips" in which all stakeholders could highlight how their interests were addressed.

The elements of the new framework proved to be collectively exhaustive. The contributions of LAAs to collaborative planning processes that were described by 60 respondents could all be classified according to the framework without exception. The mobilisation of financial, political, legal or procedural support proved an important element of the 'governance' activity. That lobbying and dissemination work was important for the implementation of new solutions and policies, especially as the LAA had no mandate or formal position to impose conditions on to the development process. Respondents indicated that this independence was a success factor in contributing to collaborative planning as the LAAs had the freedom to address new and politically sensitive problems and explore innovative solutions, unconstrained by formal political positions with participation of different stakeholders and expertise and had access to broader and new political, research, media and public networks. LAAs in which actors are more willing to leave entrenched positions are perceived to be crucial for the adaptive governance of social-ecological systems (Folke et al. 2005). Gunderson (1999) stresses that these informal, shadow networks can be incubators for new approaches to govern social ecological systems. Van Herk et al. (2011b) suggested that the connection can be made through people working together. In particular the pilot projects 'flood proof building' of 100 dwellings in Stadshavens and 'water living' of 500 dwellings in Westflank became important connectors between the LAA and the formal decision making process towards implementation.

Also, the framework proved useful, both descriptively as an analytical tool and prescriptively to guide LAA work, although the activities, threads and streams are not necessarily mutually exclusive. The overlap between activities is logical as they are mutually dependent; e.g. respondents highlighted that "many design ideas popped up whilst discussing results of (system) analysis." The difference between the threads 'images' and 'ambitions' was not always clear either. When partners discussed their view of reality (*images*), they also discussed their shared or conflicting *ambitions* towards implementation. Finally, the researchers found that new images (thread) often led to, or were created by new solutions (stream) and that negotiating ambitions (thread) involves participants and influences politics (stream). These overlaps make a strict classification difficult. However, during the validation workshop, participants stressed the practical and analytical value of the framework. After case study presentations, 87 participants were asked questions such as: compare the importance of the LAAs in: analysing problems versus developing a solution, or establishing facts compared with creating images and setting ambitions. The subsequent discussion proved insightful. Few participants stressed both possible answers, whilst roughly half of the participants mentioned one pathway (thread or stream) of a LAA to support collaborative planning, and the other half mentioned the alternative. After group discussion, participants confirmed the usefulness of all activities, all threads and all streams, and the relation between them. The framework provides a tool to LAA members to organise their joint work to contribute on all threads and streams and to recognise the importance of each activity in contributing to these. It is possible to conclude that the value of the framework is in recognising the broadness and added value of all activities and aims, not in strictly separating them.

2.8 Conclusions

Urban development, regeneration and retrofitting present windows of opportunity to reduce flood vulnerability that are often not taken advantage of. Collaborative planning is needed to integrate urban retrofitting and development processes and flood risk management. This can be achieved by a social learning framework: Learning & Action Alliance (LAA). A new framework has been developed and validated to guide the organisation of LAAs and to analyse their performance in support of collaborative planning. A LAA should be organised in terms of a number of activities: system analysis; collaborative design and governance in order to: establish facts; create images and set ambitions (*threads*) and to: address problems; develop solutions and involve participants and influence politics (*streams*). The two case studies in the Netherlands show the value of LAAs that had a decisive influence on the flood proof, urban development masterplans and related policy proposals that have been adopted. LAAs can be a governance or coordination mechanism or policy instrument (Blackmore et al., 2007; Ison & Watson, 2007) to create flood resilient urban areas by taking advantage of current redevelopment and retrofitting opportunities.

Chapter 3

Process design and management for integrated flood risk management: exploring the Multi Layer Safety approach for Dordrecht, The Netherlands

Chapter 3 Process design and management for integrated flood risk management: exploring the Multi Layer Safety approach for Dordrecht, the Netherlands

This chapter has been published as:

Van Herk, S., Zevenbergen, C., Gersonius, B., Waals H., Kelder, E.T.G. (2013) Process design and management for integrated flood risk management: exploring the Multi Level Safety approach for Dordrecht, the Netherlands. Netherlands. *Journal of Water and Climate Change*, doi: 10.2166/wcc.2013.171

Abstract

New flood risk management policies account for climate and socio-economic change by embracing a more integrated approach. Their implementation processes require: collaboration between a group of stakeholders; combining objectives and funding from various policy domains; consideration of a range of possible options at all spatial scale levels and for various time horizons. Literature provides limited guidance on how to organise a collaborative planning process to devise integrated flood risk management (IFRM) plans. This paper presents a case study where a recently developed framework for process design and management has been used and evaluates whether or not the collaborative planning process led to an IFRM plan. The case study is Dordrecht (NL) where the new multi-layer-safety (MLS) approach has been applied in the context of the Delta Programme. The Delta Programme investigates how the Netherlands can adapt to the effects of climate change. MLS comprises 3 flood safety layers to reduce flood risk: flood protection; spatial planning; and emergency response. The framework has been shown to be effective in the delivery of an IFRM plan; it has been enriched by defining the interfaces between and phasing of planning activities; and can be further improved to better guide implementation and governance activities.

3.1 Introduction

Traditionally, flood management practices in Europe have focused on predominantly defensive approaches to reduce the probability of flooding (Newman et al, 2011). However, in the past two decades major flood disasters have created the need to shift from flood protection to a more integrated approach in which flood risk is actively managed to also reduce flood impacts (White, 2010; Dawson et al. 2011). This rethinking of and change in the traditional approach is included in policies such as: the EU Flood Directive (EC, 2006); the source-pathway-receptor framework as used by the Environment Agency in England and Wales (EA, 2000); and the *multi-layer safety* approach (MLS) in the Netherlands (V&W, 2008) that has been studied for this research. The EU Flood Directive endorses an integrated approach to flood risk management (IFRM). It requires the member states to prepare IFRM plans for areas where risk is deemed significant. These should include appropriate objectives and measures that focus on the reduction of the likelihood of flooding and/or on the reduction of the potential adverse consequences. IFRM plans are meant to address all phases of the flood risk management cycle, but focus particularly on prevention, protection and preparedness (adapted from: EC, 2004):

1. *Protection*: taking measures to reduce the likelihood of floods in specific areas, such as by building flood defences or preventing high river discharges;
2. *Prevention*: using spatial planning and adaptation of buildings to prevent or reduce damage if a flood occurs;
3. *Preparedness*: improving organizational preparation for floods, such as emergency plans, risk maps, risk communication, early-warning systems, and flood insurance;
4. *Emergency relief*: providing emergency relief in the case of a flood, such as evacuating communities, erecting temporary flood defences and providing medical help;
5. *Recovery and lessons learned*: mitigating the social and economic impacts on the affected communities, and undertaking surveys to identify the conditions that led to effective mitigation and adaptation measures, or an effective emergency response to the flood event.

The Netherlands is opting for a MLS approach, which is laid out in its National Water Plan (V&W, 2008). This is a three-layered approach that goes beyond flood protection, which is its first safety layer. The two other layers are aimed at reducing the consequences of flooding by adapting the spatial layout (second safety layer) and enhancing emergency response (third safety layer), respectively. Hence, MLS explicitly calls for the consideration of measures to reduce adverse consequences and to operationalise these measures.

The implementation of these policies and subsequently the development of IFRM plans result in the need to combine objectives and funding from various policy domains; to consider a wider range of possible measures at all spatial scale levels and for various time horizons; to involve multiple disciplines and, as a result to collaborate between a group of stakeholders with various interests and means (e.g. Potter et al., 2011). Few authors (Van de Ven et al., 2011; White, 2008; Van Herk et al, 2011a) offer guidance on how to organise a collaborative planning process to devise IFRM plans. Few examples of such processes have been identified, in practice, documented and evaluated (Pahl-Wostl et al., 2007) as the implementation of truly *integrated* FRM plans is still in its infancy (Huntjens et al., 2011). White (2008) lists a range of activities and possible measures for IFRM, but does not provide ideas on how to organise these activities into a coherent process. Van de Ven et al. (2011) propose a step-wise process comprising vulnerability analysis, selection of a strategy and selection of measures, that is an idealised-type of process that so far lacks empirical evidence. Van Herk et al. (2011a) used empirical evidence to develop a framework to design and coordinate a collaborative planning process that clusters the activities described by White (2008) and Van de Ven et al. (2011) into 3 groups: system analysis; planning, design and engineering; and governance. This framework requires validation to provide practical guidance for prescriptive, rather than descriptive use. If an IFRM process is organised using this framework, does it support the production of an integrated plan?

This paper describes the process of the development of an IFRM plan that was organised from the outset following the collaborative planning framework (van Herk et al., 2011a) and aims to evaluate the process and the framework in terms of supporting the delivery of an integrated plan. An IFRM plan comprises: integration of objectives including across spatial and temporal scales (see also: Dovers, 2005; Zevenbergen et al., 2008; Rijke et al., 2012b). The research described here is based on a case study in the city of Dordrecht in the Netherlands, where an IFRM plan has been developed based on the MLS approach. MLS inherently embraces IFRM as it stimulates the combination, coordination and planning of investments in measures from the three flood safety layers. IFRM based on MLS aims to develop an integrated strategy to reduce present and future flood risk that comprises a portfolio of options. Options at multiple scale levels in the flooding system that can be implemented at various moments in time, and can be combined with objectives from other related policy domains such as transport, housing, and environment. Hence, the MLS approach and the planning framework used in Dordrecht embody generic IFRM concepts that may be used for international comparison. In practice, the case study in

Dordrecht has replication potential because it is a pilot project of the Dutch Delta Programme (Gersonius et al., 2010) that will coordinate the implementation of Dutch water Policy for the coming decades (Deltacommissie, 2008; Zevenbergen et al, 2013a; 2013b).

3.2 Evaluation framework and methods

A mixed-method case study approach has been used to analyse the collaborative planning process and evaluate it in terms of supporting the delivery of an integrated plan. First, the IFRM plan has been evaluated in terms of integration of objectives across spatial and temporal scales. Integrated outputs may align and balance *multiple objectives*. Flood risk inherently comprises multiple and sometimes competing objectives, such as reducing: individual risk (probability of death of one person); group risk (function of probability of flood event and related number of fatalities); economic risk (function of probability of flood and related direct and indirect material damage). Moreover, physical interventions to reduce flood risk need to be incorporated into spatial planning and thus need to be aligned with objectives related to e.g.: housing; nature; economics; water quality; transport; etc. to increase the political and financial feasibility of the implementation of the interventions (White, 2010; Veerbeek et al, 2012). IFRM outputs can comprise a single or multiple options from all relevant *spatial scales* in the flooding system (see also Adger et al., 2005; Zevenbergen et al., 2008). A MLS plan or strategy can comprise measures to protect, prevent and prepare e.g. individuals or communities; individual buildings or an entire city or catchment. The plans are developed considering various *time scales* in order to balance short and long-term costs and benefits and anticipate (potential) future change, such as climate (*ibid*). Various documents have been analysed including: the final plan (DPRD, 2013); intermediate products from the different planning activities (MARE, 2011); external research reports from directly involved or external researchers (e.g. Gersonius et al., 2010 and 2012; Van Herk et al., 2011b); minutes of meetings; policy documents; and media.

The collaborative planning process has been analysed following the same framework that has been used to design and coordinate the process (Van Herk et al., 2011a). The planning activities have been classified into 3 groups: system analysis; planning, design and engineering; and governance that generate knowledge to contribute to decision making (Van Buuren, 2006). System analysis generates information to analyse and address problems, for example: the areas at risk from flooding; green infrastructure mapping; ground composition; flow paths; nature and vulnerability of urban fabric; critical infrastructure (White, 2008). Planning, design and engineering aims to develop strategies and measures through e.g. workshops with multiple stakeholders and typically results in for example: land use plans; urban masterplan designs; or designs of specific elements such as levees or buildings. The governance activity supports decision making by involving participants in discussing and defining their ambitions and roles regarding a project or policy, or the problems and solutions as defined in the activities of system analysis and planning, design and engineering. Complex decision making, such as for IFRM, involves stakeholders with different views, stakes and values in interpreting and valuing information appropriately; mobilizing specific knowledge to underline viewpoints (Koppenjan & Klijn, 2004). Semi-structured individual and group interviews have been conducted with participants (n=17) to analyse the planning process and its contribution to an integrated plan. These interviews focused specifically on the collaborative process: activities conducted; knowledge they generated; and how these supported dialogue for decision-making throughout the development of the IFRM plan. The interviewees were representatives from the Ministry; Waterboard; Province; Municipality; Safety Region; and research institutes involved. The researchers were able to gather much data and gain a clear understanding of the activities and their contribution to the integrated outputs, as the researchers were themselves also participants in the collaborative planning process and thus engaged in delivery as well as observation. This is *action research* (Checkland and Holwell, 1998; Flood, 1998). The researchers designed and facilitated the collaborative planning process themselves and followed the process framework to evaluate its potential to deliver integrated outputs.

Validation interviews were conducted with participants in the process, that were not interviewed previously, to validate the influence of the collaborative process on the integrated outputs (n=6). The interviewees represented 4 government agencies and 2 private stakeholders. Additionally a survey has been conducted amongst the participants (n=42) of a workshop on multi-layered safety for the case study. Respondents were asked about the opportunities and barriers for an integrated process and to develop integrated outputs based on the MLS approach to get an understanding of the potential for wider uptake. The qualitative data obtained from the open questions have been analysed for recurrent similar statements. The respondents comprised representatives of: police; fire brigade; red cross; national, regional and local government; utility companies; construction companies; engineering firms; research institutes and universities.

3.3 Application in a case study: pilot Dordrecht

This Section describes the case study; the collaborative planning process; and evaluates qualitatively the resulting IFRM plan of MLS in Dordrecht in terms of integrated outputs.

3.3.1 Case description

The application of the process framework is here illustrated using the example of IFRM for the Island of Dordrecht, the Netherlands. Surrounded by a series of rivers and canals – the Oude Maas, Beneden Merwede, Wantij, Nieuwe Merwede, Dordtse Kil – the city of Dordrecht is located on an island. The population of Dordrecht consists of around 120,000 inhabitants. Most residential areas are located in a single polder area of about 7 ha, which is protected by a 37 km long dyke-ring. The unembanked areas along the riversides have a higher ground level and comprise: a part of the historical inner city; industrial areas; residential areas; recreational and nature areas. The Island of Dordrecht lies in the transition zone between the tidal reach and the river regime reach where the extreme water stages are influenced by both the high river runoff and storm surges from the sea. Together with the city of Rotterdam, a number of smaller cities and the surrounding agricultural and nature areas, the Island of Dordrecht is located in the Rijnmond-Drechtsteden area. One of the six regional sub-programmes of the Dutch Delta programme was set up for this area.

In the context of the Dutch Delta Programme, in 2010 the Dutch government commissioned a MLS pilot project for the Island of Dordrecht (among other) to improve flood safety in an area-oriented development processes. The goal of this pilot project was twofold: 1) to deliver enhanced knowledge and expertise to national policy; and 2) to develop a regional vision for flood safety and insight into possible measures to improve flood safety, potentially combined with other goals to be converted into a regional IFRM plan. This research focuses on the pilot's second goal by analysing how the collaborative process supported the delivery of integrated outputs in the IFRM plan from 2010 to 2013. The pilot project took place under the direction of the regional authorities (municipality, province, water board and safety region), where the national government actively participated, but did not direct. A collaborative planning process was required for the pilot project, because multiple stakeholders needed to be involved to develop and implement measures in the three safety layers. The distribution of responsibilities amongst government agencies is regulated and clearly defined for flood protection in the area. The water authority Hollandse Delta is responsible for operation and maintenance of the levee system around Dordrecht, whilst the National

Government sets related protection standards and funds the investment costs for these levees. MLS, however, requires collaboration between a wider group of stakeholders with various interests and means that are not (yet) coherently managed, funded and/or regulated. The second safety layer of spatial planning is primarily the responsibility of the Municipality of Dordrecht at a local level and of the Province of South Holland at a regional level. Additionally, many other stakeholders affect or are affected by spatial developments such as: private developers; landowners; and inhabitants. The Safety Region South-Holland-South is responsible for emergency planning in the third safety layer. The Safety Region comprises emergency services such as: the fire brigade; police; and ambulances. Also, increased preparedness of operators of critical infrastructures such as water and energy utilities and communication companies can reduce flood consequences.

For the pilot in Dordrecht stakeholders came together in a Learning & Action Alliance, a 'shadow network' that comprised voluntary participation of representatives from all of the organisations involved. A shadow network was considered necessary by the participants, as there is no blueprint to organise the IFRM process for the application of the new MLS approach. A shadow network is an independent platform for collaboration to work with alternative approaches for governing socio-ecological systems (here: flooding system) to experiment and generate alternative solutions to emerging problems (Olsson et al., 2006). The shadow network consists of representatives of relevant stakeholders with no formal authority. Consequently, it does not adhere to existing governance structures and provides a learning environment where stakeholders can leave entrenched positions to work and learn together unconstrained by formal political positions (Van Herk et al., 2011a). Such informal participatory platforms can occur even in a rigid and strongly structured administrative environment (Moellenkamp et al., 2010). Existing institutional arrangements are often not sufficient for IFRM (Huntjens et al., 2011; Wong and Brown, 2009) and the same holds true for the Netherlands and MLS. The existing procedures and instruments in the Netherlands adhere to persistent government structures and policies that were designed based on a sectoral approach for flood protection, rather than integrated approach to flood risk management (Van der Brugge et al., 2005). Hence, the first goal of the pilot was to support national policy development. Related to the second goal, to make an IFRM plan, the IFRM process comprised an initial stage of strategy development, whilst actual realisation of the plan, including the delivery of urban (re)development and infrastructure development projects, if at all, is foreseen as happening over the coming decades. These aspirations required flexible network structures and governance that allowed for learning.

3.3.2 Integrated outputs

The question addressed here is if the IFRM plan for Dordrecht comprises integrated outputs: in objectives, and across spatial and temporal scales.

Integration across *spatial scales* is central to the MLS strategy and IFRM plan. The flood hazard can only be reduced at a national level, as it can be caused *inter alia* by a storm surge at the North Sea (downstream) and peak discharge in the rivers (upstream). The flood hazards were taken as a starting point for the IFRM plan for Dordrecht. On the regional scale, in this case the Island of Dordrecht that includes the dyke ring area and the unembanked areas, the flooding system has been designed to be 'self-reliant' in case of a flood. Evacuation of the entire population from the island was not considered feasible, or effective (see: sub-section on system analysis). Self-reliance was, therefore, selected as a central concept that aimed to strengthen and align the capacity of organizations, communities and individuals each operating at different spatial scales, to cope with and recover from an extreme flood event (i.e. above the current protection standard). Measures that contribute to increased self-reliance comprise the realization of a 'delta dyke' and compartmentalizing the dyke ring area into three subdivisions by using the existing regional dykes (see Fig. 3.1). The northeast part of the dyke ring is to be made virtually unbreachable (100 times lower probability of failure than for the current design standard) by heightening and/or widening the dyke sections. This allows for overtopping of the levees with flood water, rather than breaching, and thus reduces the potential consequences of a flood event. The three compartments allow for a differentiation of flood risk and specific strategies for each compartment. The south 'compartment 3' consists mainly of agricultural land where the consequences of a flood would be lower than in the 2 north compartments. Compartment 3 will have a lower protection standard and further urban development there is to be avoided. The northwest 'compartment 2', that has a higher individual flood risk, could be evacuated to the northeast 'compartment 1' in case it is threatened by a levee breach. These measures also provide opportunities for effective measures at the local, neighbourhood scale that are proposed to further reduce the consequences of flooding and enable the functioning of the socio-economic system during and after a flood event, such as: individual protection of critical infrastructure networks (e.g., energy supply) and nodes (e.g., hospitals); creation of safe havens or shelters; elevated access and egress routes within and between compartments and neighbourhoods; flood proof planning and building of public spaces and buildings; and improved risk and crisis communication.

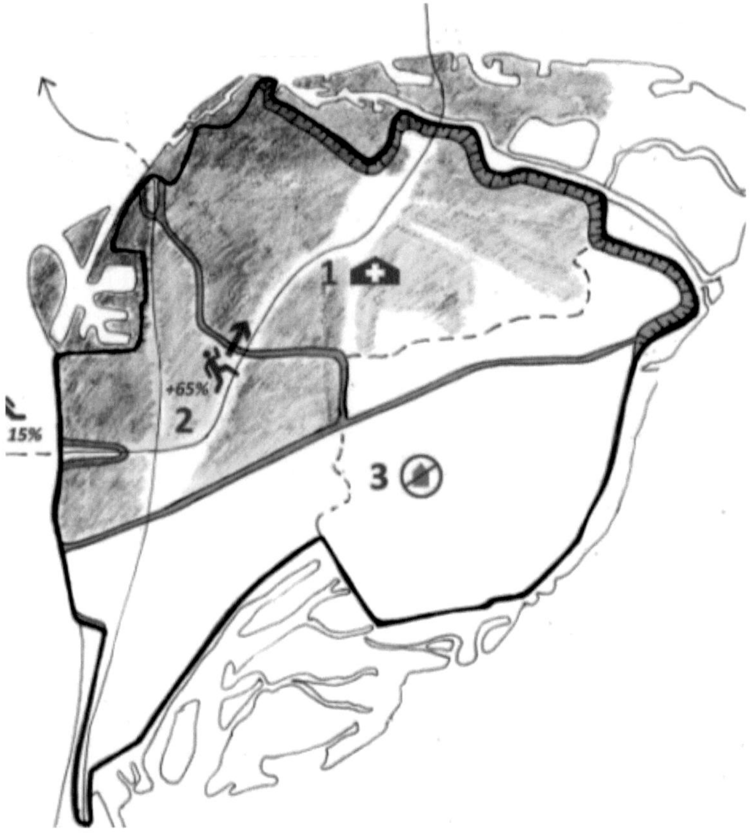

Figure 3.1. Illustration of the IFRM plan for the Island of Dordrecht.
The fine black line indicates the dyke ring, with a virtually unbreachable delta dyke in the North East that merely allows for overtopping of floodwater; and the compartmentalizing of the dyke ring area into three subdivisions by secondary dykes (grey lines) allowing for differentiation of flood risk and specific strategies. No further development is allowed in compartment 3. 65% of Inhabitants of compartment 2 can be evacuated to compartment 1 prior to a flood event. And shelters are to be created in compartment 1

Integration across temporal scales is inherent to the self-reliance strategy and is included in the FRM plan in two ways: to deal with various potential futures and for a flexible implementation strategy over time. Firstly, the plan is robust and flexible for various potential long-term futures, taking a range of climate and socio-economic scenarios into account. As self-reliance increases the capacity to cope with and recover from an extreme flood event, it is less dependent on a design standard for the dyke protection system that is designed to handle an expected return period and future climate scenarios. For example, the entire dyke ring, with the exception of dyke section 13 (Voorstraat), has been designed with a residual height in order to accommodate climate change and sea level rise under the projected KNMI climate change scenarios for 2100 (Van den Hurk et al.,

2006). The Voorstraat, which is the primary dyke through the historical inner city, is not high enough to withstand higher water levels than those accounted for by the current design standard. This dyke section is difficult to heighten without affecting cultural heritage buildings (Gersonius et al., 2012). However, it is virtually unbreakable because of its very flat inner slope and stone cover. For this dyke section the IFRM plan proposes the construction of dry-canals behind the dyke to accommodate floodwater in case overtopping and overflow occurs more frequently in the future as expected due to climate change. Secondly, the investment planning for the proposed measures is optimised over time to reduce investment costs by exploiting the many opportunities for mainstreaming and integrating flood risk management into sectoral policies, planning frameworks, and 'normal' investment and renovation cycles. The construction of the delta dyke is to be combined with already envisaged dyke reinforcement projects, such as for the dyke section at Kop van het Land in the East. Several schools have been selected to serve as shelters to receive evacuees in times of emergency. When regular renovations to these schools are scheduled, they will be adapted to additional requirements for a shelter where opportune. The (Wieldrechtste) Zeedijk, a regional dyke separating the urbanised areas in the north from agricultural and natural areas in the south, is to be strengthened to add additional protection to the urbanised north. This strengthening is to be combined with envisaged nature development along the dyke that will be designed to break flood waves, in addition to enriching the landscape. Incremental investments over time, as proposed by the IFRM plan, provide the flexibility to adjust the implementation and timing of measures based on new insights, e.g. from updated climate and socio-economic scenarios.

Integration in objectives is a direct consequence of the three-layered MLS approach that aims for multiple flood risk objectives by reducing probability and consequences of floods, but has also been achieved by incorporating socio-economic objectives. The dyke reinforcement planned in the East of the Island is to be combined with nature development on the riverside of the dyke. The dyke reinforcement in the West supports the redevelopment of an industrial area that is intersected by the dyke, providing a safe elevated transport route. The Stadswerven, an unembanked old industrial area in the North of the island is being redeveloped into a residential and commercial area. Spatial and financial objectives have prevailed in the related master planning, but flood risk was used as a design variable. The masterplan includes lower lying, attractive water-rich areas to appeal to potential house buyers. The public spaces and dwellings are flood-proofed by elevated construction, dry-proof building design or by selection of materials, pavements and vegetation. Transport infrastructures are also elevated so that they can serve as escape routes for evacuation. The construction of dry-canals to guide floodwater overtopping De Voorstraat dyke, requires integration of these flows into planning and maintenance of public roads in the embanked polder area. Regional businesses have been invited to explore business opportunities related to the implementation of the IFRM plan in Dordrecht so as to support economic development and also any exploitation potential internationally. Examples include: advanced sensors and ICT systems for flood monitoring and management; engineering and construction of delta dykes and flood proof dwellings.

3.3.3 Collaborative planning process

In this section the collaborative planning process will be described using: (i) system analysis; (ii) planning design and engineering; and (iii) governance as given in the framework of Van Herk et al. (2011a).

*System analysis**

Extant analysis results from continuous monitoring of system performance were inventoried. The Dutch Environmental Assessment Agency had conducted a risk zoning analysis for Dordrecht and surrounding areas to classify risk areas in terms of maximum water depth and lead-time; variables with the greatest influence on casualties in the Netherlands (Pieterse et al., 2009). The dyke ring area of Dordrecht is at high-risk, where flooding would either occur that would be 'deep and slow' or 'deep and fast'. These insights brought attention to the problem of residual risk beyond the design standard: what if a dyke breaches? The east side of the island was found to be one of the most dangerous breaching locations. From flood modelling, such a flood has been estimated to cause damage of 4.6 Billion Euro (approximately one third of the total assets in Dordrecht), cause 566 casualties and would affect 96% of the population living in the embanked areas. As a result, the ambition was set to avoid a dyke breach at this location. This stimulated the exploration of alternative options that led to the proposals for the 'unbreacheable' delta dyke and the compartmentalizing of the dyke ring area into subdivisions (as shown in Figure 3.1). Evacuation of the entire population from the island was not considered feasible, or effective to reduce casualties. Because of a short warning time (less than 24 hours) only 15% of the population could be evacuated and adjacent areas would potentially also be flooded, with the four main egress routes becoming inaccessible due to the storm conditions. This conclusion prompted the development of the self-reliant strategy. Flood

* see Gersonius et al. (2013) for an elaborate presentation of the methods applied for, and results of all analyses

modelling, later confirmed by an analysis of ground elevation maps, led to the understanding that the unembanked areas are situated on higher ground (on average around 3 meters higher than the embanked areas) and thus less vulnerable for flooding than the embanked areas as flood levels would be lower. Subsequently, the options have been explored to use the unembanked areas as safe havens and to host shelters and use them for the provision of critical infrastructure functions as part of the self-reliant strategy. The Statutory Assessment for the Island of Dordrecht found that 28% of its flood defences do not comply with the legal standard and require reinforcement before 2015 (PZH, 2011), whilst the Voorstraat dyke was expected to be found to be inadequate during the next Assessment round. The reinforcement projects required (I&M, 2011) were later inventoried to determine if they could provide opportunities to combine investments for cost-effective implementation of the IFRM strategy. Many other analyses were also conducted, e.g.: vulnerability assessments of the cultural heritage buildings in the unembanked areas; of the critical infrastructure; and of the access and egress routes.

In a second phase of analyses, the various measures that had been developed were assessed and compared to support dialogue on political feasibility and funding. These measures were compared in terms of their effects on the Expected Annual Damage and Expected Annual Number of Casualties. Also, for the combined self-reliance strategy comprising a set of measures, the graduality of flood risk was studied for different return periods (see Fig. 3.2) following De Bruijn (2005). Graduality refers to the relative increase in losses with increasingly severe flood events. This provided insight into the effectiveness of the strategy for various return periods, and thus indirectly for alternative climate change scenarios that would decrease return periods, i.e. flood events would occur more frequently. Structural assessment of protection measures (Den Hengst, 2012) and prevention measures (Blom et al., 2012) that supported the engineering work was carried out in the EU FP7 Floodprobe project (FloodProBE, 2013). A new evacuation strategy has been evaluated for the 'evacuation fraction' that has been increased from 15% of the population leaving the island to 80%; because 65% of the population can evacuate from compartments 2 and 3 to the safer compartment 1 in the North East (Figure 3.1). Also story lines, a narrative of a series of events and actions during a flood event, have been developed (Lips, 2012) to support the organisational preparation and improve the emergency response (third flood safety layer).

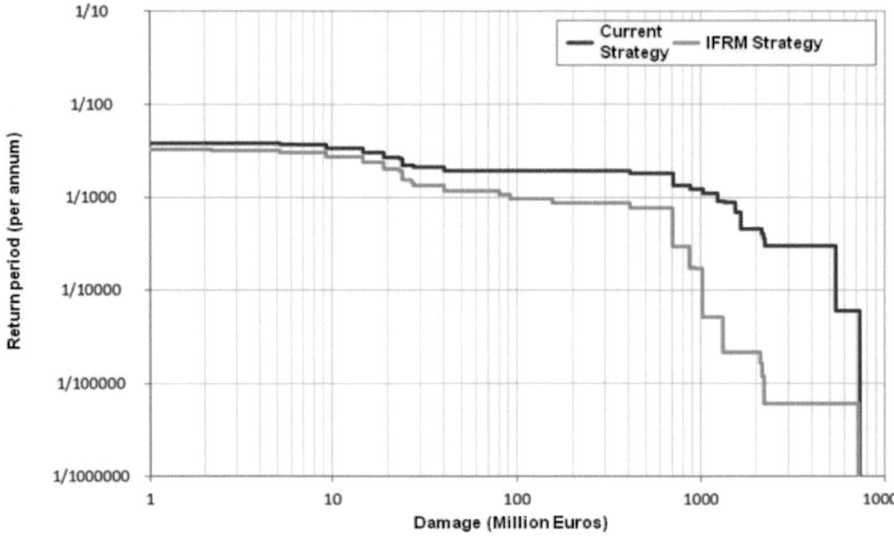

Figure 3.2. Graduality of flood risk in terms of damage (Millions of Euros) for the Island of Dordrecht, with the current FRM strategy (in black) and the IFRM strategy (in grey)

Planning, Design and Engineering

This activity started with an inventory of possible measures to reduce flood risk based on existing literature (e.g. Van de Ven et al., 2009) and examples from other cities intended to inspire participants. Subsequently a stakeholder workshop was held with 42 representatives of: police; fire brigade; red cross; national, regional and local government; utility companies; construction companies; engineering firms; research institutes and universities. Initially and without prior instruction, participants started discussing the existing levee system and possible improvements framed in the context of the traditional approach of flood protection. Later, the participants brainstormed options from the second and third safety layers using the knowledge from their respective domains and considering their means and current practice. This led to increased understanding of each other's means and interests and an inventory of constraints, needs and research questions. E.g. the representative of the gas utility company stressed that water pressure from floodwater can damage the underlying gas lines. This insight might have to be converted into a design requirement for new developments in the unembanked areas. Professionals from the emergency services listed the various water levels that would impede circulation of their vehicles, such as ambulances, or army vehicles and would thus hamper emergency response. Urban planners made a visual inspection of the dyke ring to inventory the characteristics of various dyke sections and their surroundings. This comprised an analysis of available physical space adjacent to dykes that could possibly be used for the widening of their foundations required for dyke strengthening. They found that certain dyke sections already have characteristics of a delta-dyke, because the ground level of adjacent land was as high as the dyke's crest height and thus added strength to the dyke. Critical infrastructure providers highlighted that water supply, energy plants and chemical factories are based in the unembanked areas on higher ground, giving rise to the idea that if properly protected, they could serve the population during a flood event and thus enable the self-reliant strategy. The police and the fire brigades realised that they have their headquarters in a flood vulnerable area and that their command centre is at the lowest level of that building. If flooded, communication could be hampered, jeopardising the emergency response operations during a flood event. Future renovation or newly planned building of the headquarters and command centre can address these vulnerabilities.

Different options were combined in an IFRM plan using the new insights as regards their costs and effectiveness (system analysis) and potential political support amongst stakeholders (governance). The former supported the idea that a combination of options from various scale levels of the flooding system can effectively deliver upon multiple objectives. A holistic strategy was proposed, consisting of: a protection strategy with a delta dyke and compartmentalisation dykes; a self-reliant strategy with flood-proofed critical infrastructure, shelters and area-specific evacuation; and a development strategy coupling FRM measures with spatial measures over time. The latter is crucially influenced by the mandates of different authorities and the available funding. E.g. the transformation of a dyke section to a delta dyke proved not viable under the current governance framework, because there is no national assessment framework for the prioritisation of dyke sections and thus for funding and implementation of delta dykes. Coupling with existing (re)development and maintenance works is necessary to reduce costs, and, therefore, planners looked for such opportunities. As a result the self-resilient strategy can only be implemented over a long period of time, with some redevelopment projects in the unembanked areas not being foreseen for the next 20 years. This poses an important challenge to decision-making and governance regarding the political feasibility to sustain ambitions and commitments for a long-term strategy with multiple objectives for various stakeholders each or any one of which could have a shorter time horizon.

Governance

A stakeholder analysis was conducted prior to engaging the relevant stakeholders. It followed the approach of Klijn & Van Twist (2008) and Koppenjan & Klijn (2004]: identify involved stakeholders; their perceptions; and their positions and interdependencies. Inspired by the MLS concept, stakeholders from the second and third flood safety layer were involved for the first time in discussing water safety with traditionally dominant stakeholders from the first layer. During the stakeholder workshop they got to know each other and discussed their roles and ambitions. As one survey respondent stated: ¨we know of the existence of each organisation, but don't know the people, let alone work together.¨ And another: ¨No organisation has a coordinating or steering role over the three layers and every organisation is fighting for its role......At a national level the three safety layers are decoupled in 3 ministries. Now you can still work on flood safety from your own policy sector without coordinating.¨ During the workshop they sought for synergies by combining their means to explore integrated strategies and combinations of options. After further design work and analyses, the Municipality, Water Board, Safety Region and National Ministry, discussed possible implementation paths and investment strategies for the IFRM plan. However, to date no decisions have been made on the investments for and implementation of the IFRM plan, because of barriers imposed by existing government structures and policies. Current national regulation merely sets design standards for levee systems based on return periods and thus national funding is merely provided for defence structures. Implementation of measures from the other safety layers is to be funded by regional stakeholders and without legal obligation (Kolen et al., 2010). This focused discussions on how to combine objectives and funding from other policy domains. In parallel, separate workshops were organised to discuss the need for policy change. An exemplary discussion was held on the solution proposed for the Voorstraat dyke with dry-canals to accommodate flood water in case of overtopping. A shift of responsibilities and budgets is required to fund, build and operate the dry-canals. A political discussion document has been produced by policy advisors, but is yet to be discussed by elected politicians. The policy advisors indicated in an interview that the Waterboard is interested in the approval of the dyke section without heightening, but considers that the decision for funding the construction and maintenance of dry-canals can be postponed. This poses a risk to the implementation of the measure. Similarly, for the dyke reinforcement project Kop van het Land, the project manager and responsible politicians of Waterboard, Municipality and National Government have been presented the opportunity to convert that dyke section in a delta-dyke, but no agreement has been reached yet on the funding of the additional costs that are not mandatory under current regulation. The political feasibility of the delta-dyke upgrade is decreasing, as the preparation continues of the already planned and funded dyke reinforcement project that is less ambitious. Vulnerability analysis showed that the upgrade of this dyke segment is a prerequisite for a self-reliant strategy, because the consequences of a breach here cannot be mitigated with the proposed measures on the second and third flood safety layers.

3.4 Discussion on the process framework and its general applicability

The contribution of the collaborative process to an IFRM plan is discussed below, and the process framework has been enriched by the analysis of the interaction between activities and the phasing of the activities to develop an IFRM plan.

3.4.1 Contribution of the collaborative planning process to integrated outputs

Based on the case study analysis it can be concluded that the collaborative planning process led to the development of an IFRM plan. The IFRM plan that was developed is integrated across spatial and time scales and aligned multiple objectives; and the three activity groups in the collaborative planning process have contributed to the integration in the IFRM plan. For example 'system analysis' assessed the performance of IFRM for different objectives (e.g. individual risk and economic risk) and for various climate change scenarios. Planning, design and engineering devised options for various scale levels such as: upgrading the dyke-ring around the Island and converting individual buildings into shelters for evacuees. 'Governance' involved a wide range of stakeholders ranging from: private developers that can flood proof a to-be-build residential neighbourhood whilst improving its value on property market, to the national government to discuss safety standards and funding for the dyke system. The contributions of the planning activities to the integration elements of the IFRM plan have not been straightforward and depended on the involved practitioners and the coordination of their work. E.g. the analysts were told to assess strategies on more criteria than the reduction of potential damage and casualties of different flood events. Urban planners themselves looked to combine flood measures with investment plans from other policy domains. Table 3.1 summarises how each activity group can contribute to integration, which provides an evaluation framework and guidance to project managers of future IFRM projects.

Much literature is available on how specific activities contribute to the various elements of integration (in objectives and across spatial and temporal scales), but not from a holistic process-oriented perspective as analysed in this research. Some examples are given here. Research on climate change scenario analysis abounds to consider various temporal scales in 'system analysis', such as related to the Assessment Reports of the Intergovernmental Panel on Climate Change (IPCC). In terms of 'planning, design and engineering' Gersonius (2012) provides methods to devise strategies and portfolios of measures to be implemented over time horizons up to 50 to 100 years. Also many concepts have been proposed to 'design' with multiple objectives, such as 'Building with nature', combining coastal protection with land use planning through land reclamation using sand from the sea for dunes and beaches (Waterman, 2008). The concept of 'Green infrastructure' (Sandström, 2002) aims to utilise green systems as much as possible in urban space, that if done properly can help reduce and manage flood risk (Gill et al., 2007; Tzoulas et al., 2007). 'River widening' that aims to increase river discharge capacity provides opportunities to increase spatial quality (Rijke et al., 2012b; Opperman et al., 2009). 'Governance' literature is emerging to integrate water and flood management across various spatial scale levels under the concept of multi-level-governance (e.g. Lundqvist, 2004; Kern et al., 2008; Pahl-Wostl, 2009).

Table 3.1. Evaluation framework for the contribution of a collaborative process to an IFRM plan

Activities	Contribution to elements of integration: **multiple objectives, spatial scales and time scales**
System analysis	Analyse the performance of the flooding system and proposed strategies at all spatial scale levels (e.g. dyke ring, individual buildings), on various objectives (e.g. flood risk, cost-benefit); and for various time scales (e.g. under various climate change scenarios or investment scenarios)
Planning, design & engineering	Explore entire scope of flooding system for options from all spatial scales to reduce flood risk on various time scales; combine options from various domains delivering on multiple objectives; with various investment planning horizons (time scales).
Governance	Involve stakeholders from all spatial scales in the flooding system to support decision making with their interests (objectives) and their means to be applied or invested over various time scales.

The validation interviews indicated that the MLS concept itself was an important catalyst to set up an integrated process to deliver an IFRM plan. The MLS approach implicitly broadened the objective –from reducing flood probability to (including) reducing potential consequences of flood events- and explicitly spanned the boundaries and design freedom to develop alternative solutions with more integrative elements. The interviews, however, also confirmed that the collaborative planning process based on the framework has been instrumental to deliver the IFRM plan. As one interviewee appositely stated: "the deliberate use of results (of analysis) and visualisations (of designs) supported stakeholder dialogue towards integrated solutions". Initially, however, stakeholders started working with a traditional approach. The first results of system analysis presented the potential consequences of dyke breaches that directed initial design work towards options to upgrade the dyke system. Interviewees stressed that this has not been because of the process design, but rather is a result of the features of the flooding system that is technically locked-in to flood protection measures. Later during the process, options from the second and third flood safety layer have been incorporated in the IFRM plan.

3.4.2 Enriched framework explaining interaction between activities

It has also been validated in this case study that the three activities are mutually depended to develop an IFRM plan. For example: insights that were gained through 'system analysis' on the vulnerability of the Island of Dordrecht led to the reframing of the problem amongst stakeholders ('governance') that the potential consequences, rather than the probability of flooding, need to be decreased. This provided a basis to develop new strategy in 'planning, design & engineering', namely to become self-reliant, comprising measures such as: an overtoppable dyke ring, compartments to manage overtopped flood water and shelters. The rationale of Van Herk et al. (2011a) behind the classification of the activities is related to the different knowledge that is generated and its contribution to decision-making in planning. System analysis mainly aims to establish facts, coherent and not contradictory knowledge to reduce uncertainty, and to analyse, define or reframe and address problems. Planning, design and engineering mainly aims to create images: frame reflection in which parties identify their view of reality and discuss it, look for images or meanings that they share, and create renewed and more creative images as a result of the interaction; and to develop, analyse, discuss and propose solutions and strategies. The governance activity involves participants in discussing and defining their ambitions to create socially construed knowledge and discussing the ambitions and roles of the stakeholders towards implementation. Based on this research it can be concluded that the interaction and overlap between the activities relates to the exchange of knowledge between them. The interaction between the activities is a process of 'convergence' in the planning process where the options are reduced by feasibility checks ('system analysis' checks on performance; 'design' on technical feasibility; and 'governance' on political feasibility), rather than the process of 'divergence' through the knowledge generation in each separate activity that explores the possible options. Table 3.2 provides a generalised summary of the interaction between the activities through knowledge exchange.

Table 3.2. Interaction between the activities through knowledge exchange

Activities	Interaction between activities through knowledge exchange
System analysis	Contribute with new *facts* on the system performance and the effectiveness of the IFRM plan and its underlying strategy and constituent measures. These insights can change *perceptions* and support problem (re)framing and stimulate discussions on *ambitions* and the feasibility of various strategies and measures. The activity involves participants for analysis and interpretation.
Planning, design & engineering	Involve multiple stakeholders in planning, design and engineering to develop new solutions with their means and expertise, creating new *perceptions* on possible strategies and measures that stimulate discussion on their *ambitions*; and formulate research questions / information requirements to guide system analysis.
Governance	Involve all participants and discuss their *ambitions*. Explore solutions from the means and expertise of all stakeholders and discuss these based on their performance. Performance analysis is supported based on objectives and information provided by stakeholders.

3.4.3 Enriched framework by including phasing

The analysis of the collaborative planning process shows a phasing of sub-activities that was not prescribed by the planning framework. The observed phasing relates to the process of divergence and convergence explained by the interaction between activities. Based on the above the process framework to design and manage collaborative processes can be further extended by introducing the interfaces between activities and the phases. In Figure 3.3 the process is illustrated; with circles representing the three activity types of the framework (system analysis; planning, design and engineering; and governance); and 9 sub-activities that have been conducted in 3 different phases, either within or on the interface of the 3 activity types. In the first phase (phase 1) of 'divergence': the (flooding) system performance is analysed, e.g. in terms of the probability and potential consequences of flood events; stakeholders are brought together and set their objectives; and different strategies, options or measures to intervene in the flooding system are explored. Bringing the outcomes of these series of activities together on the interfaces between activity types leads to a new phase (phase 2) of: problem (re)framing or joint goal-setting based on the discussion of the system performance and different objectives; to discuss strategies and options based on the combined means and objectives of all stakeholders; to assess the performance of identified strategies and options, also using the objectives discussed. Ultimately (phase 3) the options are to be combined into an IFRM plan comprising a portfolio of measures and an investment and implementation plan; that is subject to formal decision-making processes between the stakeholders and within the democratically representative bodies of governmental organisations involved. Also the performance of the IFRM plan is to be analysed, which inherently is a feedback loop to the continuous monitoring of system's performance (phase 1) that will change with the implementation of the IFRM plan and due to exogenous factors (e.g. Milly et al., 2008). The set of sub-activities in the first phase can be conducted in parallel, as they are independent. The sub-activities in phase 2 and 3 have feedback loops, just as there can be feedback loops and iterations between the phases. Hence, the enriched framework as presented in Figure 3.3 can provide guidance for the design and coordination of the collaborative planning process to develop an IFRM plan, but is not a blueprint process design. Neither does the framework guide further planning phases towards implementation.

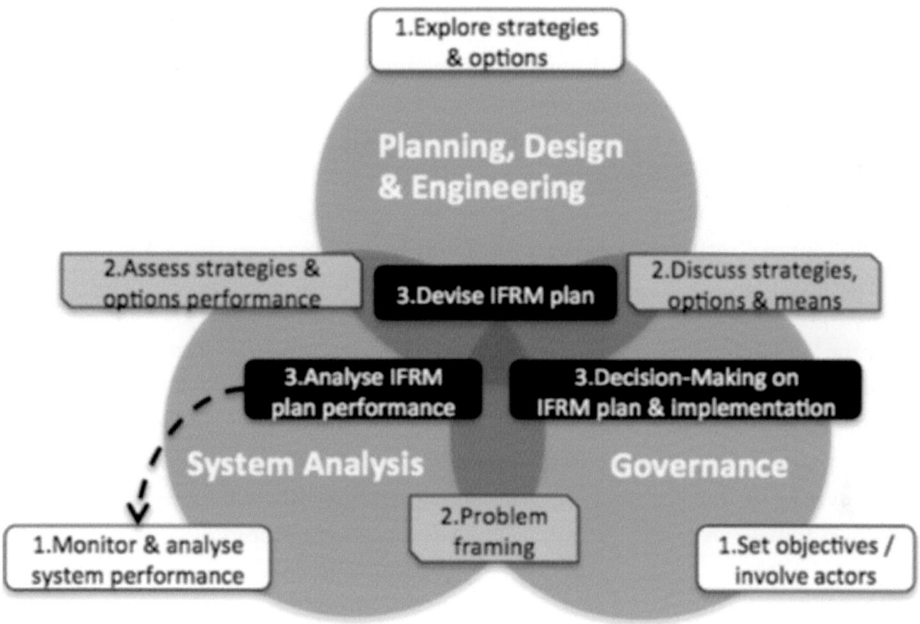

Figure 3.3. Adapted framework for collaborative planning processes to deliver IFRM plans focusing on interfaces between activities

3.4.4 Flaws in the framework: governance & implementation

The collaborative planning process in the case study also revealed some flaws in the process and the framework. All 6 participants interviewed for validation stressed the importance of the shadow network and the possibility to freely discuss problems and solutions, despite their organisation's formal position or regulatory constraints. The learning environment attracted a wide range of participants, but some key stakeholders were missing. Project managers of construction and maintenance projects and politicians overseeing those projects and their funding were only involved at the end of the process. As a result no formal agreements have been made to date and thus is the implementation of the IFRM plan uncertain. Provan & Kenis (2008) stressed the potential tension in networks related to internal and external legitimacy. Whilst the shadow network was effective in developing an IFRM, its implementation is likely to be hampered by a lack of external legitimacy in the legislative and executive bodies of government organisations involved. The implementation of the IFRM can also be hampered as current regulation and funding schemes do not incentivise the incorporation of measures from the second and third flood safety layers (Kolen et al., 2010). The challenges towards implementation of the IFRM plan posed by both the existing regulatory framework and the legitimacy tensions of the network point towards possible further improvements of the process framework presented. Namely the governance activity can comprise different aspects (as adopted from Voß and Bornemann, 2011): 'policy' (discussing problems and solutions); 'polity' (rules and structure); and 'politics' (interaction and process). Using these definitions, the case study focused mostly on policy and much less on polity and politics. In other words, governance activities are not only to discuss and define joint ambitions, but also to explore mandates and willingness to commit or combine resources (e.g. funding). Future research is needed on if and how to manage these 3 governance activities explicitly and to define output indicators for them.

Also the collaborative planning framework does not differentiate between development phases: initiation; design; construction; and operation and maintenance. The case study can be placed in the initiation or design phase when policy options have been drafted. Barriers towards implementation can be the consequence of not organizing for activities that involve stakeholders and objectives that seem more relevant in later development phases.

3.4.5 Transferability of framework and lessons:

This case study research in itself comprises the validation and enrichment of the framework that has been developed previously. Moreover, following the research presented in this paper, the process framework is already being applied prescriptively (Koukoui et al., 2013) and descriptively (Van Herk et al., 2012a) to two other case studies in the Netherlands. Initial results show that the framework is simple to apply as it provides the overarching structure of project activities, but can comprise different work tasks. Further validation and international comparison between case studies are necessary and recommended. This is beyond the scope of the research presented in this paper, but some points of attention are presented here.

Transferring best practices to other countries is likely to be a major challenge as each country has its unique geographical, cultural, institutional, and socio-economic features requiring customized approaches (Huntjens et al., 2011; Zevenbergen et al., 2013a). In this case study, the existing flooding system that relies on flood protection has been a cornerstone to the IFRM plan. The incumbent cultures and approaches initially influenced the collaborative planning process. The process framework can accommodate a range of options and approaches, but it is unlikely that it alone supports the delivery of integrated outputs without an integrated concept such as MLS. Hence, we hypothesise that an IFRM process requires an integrated concept and/or multiple objectives. Furthermore, validation interviews indicated the importance of the Learning and Action Alliance or shadow network to explore options and the potential of the MLS approach, unhampered by existing regulations and mandates. In many countries, incumbent institutional structures, regulation and policy, cultures and approaches in planning and flood risk management hamper IFRM and require a transition or regime change (e.g. Newman et al., 2011; van der Brugge et al, 2005). We hypothesise that the organizational structures that govern the collaborative process are to be fit-for-purpose and fit-for-context, recognizing the state and transition of the regime.

3.5 Conclusion

New flood risk management policies account for climate and socio-economic change by embracing a more integrated approach. Their implementation processes require: collaboration between a group of stakeholders; combining objectives and funding from various policy domains; consideration of a range of possible options at all spatial scale levels and for various time horizons. Literature provides limited guidance on how to organise a collaborative planning process to devise integrated flood risk management (IFRM) plans. The collaboration process framework presented in this paper has been found to be instrumental in delivering an IFRM plan in the case study Dordrecht. It comprises three types of activities: system analysis; planning, design and engineering; and governance. Each activity as proposed in the framework has contributed to integration by exploring possible options across spatial and temporal scales and various policy domains. They can be conducted in parallel, but are mutually enriching through feedback loops and iterations. Knowledge exchange between the activities comprised feasibility checks that furthered the planning process into subsequent phases. The validation interviews indicated that the multi layer safety (MLS) approach itself, comprising flood protection, spatial planning and emergency response, was an important catalyst in the setting up of an innovative, integrative process to deliver an IFRM plan. The MLS approach implicitly broadened the objective –from flood probability to flood risk including reducing potential consequences of flood events- and explicitly spanned the design freedom to develop alternative solutions with more integrative elements.

The collaborative process delivered an IFRM plan, but its implementation is uncertain. Legislative and executive stakeholders have not yet adopted the plan, because they were only involved at the end of the process. Moreover, current regulation and funding schemes do not incentivise the plan's implementation. These challenges point towards possible further improvements of the process framework that has been developed and validated. We recommend further research into the management of governance activities and the definition of evaluation framework for them. The governance activities are not only to engage a broad range of stakeholders to define ambitions and select strategies, but also to explore their legal mandates and willingness to commit or combine resources. A balance is to be found between granting more freedom to explore new approaches such as MLS to develop IFRM plans, and stimulating the implementation of IFRM plans by early involvement of decision makers and adherence to incumbent policy and regulation.

Many countries are adopting policies such as MLS that focus on the reduction of the likelihood of flooding and on the reduction of the potential adverse consequences and that embrace uncertainty regarding future scenarios. Hence, there is a growing need to share information and best practices in the field of integrated flood risk management across countries. This also holds true for conceptual frameworks and approaches that provide guidance on how to organize the process design and management to develop IFRM plans. The process framework presented in this paper will therefore have international relevance as it will help planners, engineers and decision makers to better understand the requirements to shape the underlying planning collaborative process. It provides the overarching structure of and relation between project activities that can comprise different work tasks. It presents how the activities can contribute to integration of multiple objectives and across spatial and temporal scales. Further research is recommended for further validation and international comparison between case studies.

Chapter 4

Attributes for integrated Flood Risk Management projects; case study Room for the River

Chapter 4 Attributes for integrated Flood Risk Management projects; case study Room for the River

This chapter is adapted from:

Van Herk, S., Rijke, J., Zevenbergen, C., Ashley, R., (under review) Attributes for integrated Flood Risk Management projects; case study Room for the River. Under review at: *International journal of River Basin Management*

Abstract

Integrated Flood Risk Management (IFRM) is advocated as an approach to develop and implement measures to reduce flood risk by collaboration between multiple disciplines and a group of stakeholders with various interests and means; to combine objectives and funding from different policy domains; to consider a range of possible options at all spatial scale levels and for various time horizons. Many new projects embrace IFRM, but literature provides limited guidance on how to organise for IFRM and few examples of IFRM projects have been identified, documented and evaluated. This paper describes research that has devised attributes of effective investment projects in developing and implementing IFRM based on an elaborate multi-method case study analysis of the Dutch Room for the River (RftR) programme. Two types of attributes emerged: those that directly foster IFRM processes and outputs and those that enable the implementation of IFRM projects in practice. These attributes have been applied simultaneously from the outset in RftR and have been mutually reinforcing during the delivery of RftR. Based on an initial validation from literature on IFRM, the attributes are hypothesised to be widely applicable. They can be instrumental in designing future research on IFRM and support the development of integrated plans as well as their implementation in practice through IFRM projects.

4.1 Introduction

Traditionally, flood management practices in Europe have focused on predominantly engineering approaches to reduce the probability of flooding (Newman et al, 2011). However, in the past two decades major flood disasters have created a shift from flood protection to a more integrated approach in which flood risk is actively managed to also reduce flood impacts (White, 2010; Dawson et al. 2011). An integrated approach to flood risk management set within urban and land use planning processes is now seen as an effective way of minimising flood risk (Yovel, 2013), although this has not always been recognised in practice and empirical guidance for implementation is still lacking (e.g. Watson et al., 2011; DCLG, 2012; van Herk et al, 2011a; Scott et al, 2013). The 'integrationist agenda' as described by Medema et al. (2008) gives an idea of how broad its scope can be and implicitly shows the inherent complexity related to organising for integration: the integrated and coordinated management of water and land as a means of balancing resource protection while simultaneously meeting social and ecological needs and promoting economic development.

Integrated flood risk management (IFRM) can be considered a fundamental component of integrated water management (IWM). The theoretical development of these domains has been mutually supportive up to the point that definitions of IFRM and IWM are interchanged (e.g. Wolsink, 2006). This paper uses the definition of Thomas and Durham (2003) for IWM that is directly applicable to IFRM: "..an approach to water <flood> management that recognises its multidimensional character—time, space, multidiscipline and stakeholders—and the necessity to address, embrace and relate these dimensions holistically…." For the research described here, this definition of IFRM has been specified for: the output of projects (integration in measures or strategies implemented); and the collaborative process to develop these integrated outputs. On integrated output, IFRM comprises integration of objectives, spatial scales and temporal scales (see also: Dovers, 2005; Zevenbergen et al., 2008; Rijke et al., 2012b). Integrated outputs align and balance *multiple objectives*. Physical interventions to reduce flood risk (that can be operationalized through various objectives such as: individual risk; group risk; economic risk (Van Herk et al, 2013a)) need to be incorporated into spatial planning and thus need to be aligned with objectives related to e.g.: housing; nature; economics; water quality; transport; etc to increase the political and financial feasibility of its implementation (White, 2010; Veerbeek et al, 2012). IFRM outputs can comprise a single or multiple options from all relevant *spatial scales* in the flooding system to protect, prevent and / or prepare e.g. individuals or communities; individual buildings or an entire city or catchment (see also

Adger et al., 2005; Zevenbergen et al., 2008). The IFRM plans are typically developed considering various *time scales* in order to balance short and long term costs and benefits and anticipate (potential) future change such as climate (*ibid*). On integrated process, the definition of IFRM inherently comprises *collaboration* between multiple stakeholders and multiple disciplines (e.g. Potter et al., 2011). For this research, IFRM has been defined as an approach to develop and implement measures to reduce flood risk by collaboration between multiple disciplines; by a group of stakeholders with various interests and means; to combine objectives and funding from different policy domains; to consider a range of possible options at all spatial scale levels and for various time horizons.

No exhaustive, evidence-based guidance has been found to organise for IFRM. Only a few authors (Van de Ven et al., 2011; White, 2008; Van Herk et al, 2011a) have offered guidance on how an integrated process can be organised. The Global Water Partnership provides guidance (e.g. GWP-TAC, 2004) to successfully implement IFRM, but Medema et al. (2008) show that the governance conditions and process steps proposed by the GWP are not sufficiently detailed, are non-specific and untestable. Examples of such conditions and steps range from: having an enabling legislative and policy environment; appropriate institutional framework; to preparing a management strategy to be implemented and monitored. Logically, it is difficult to present conditions that are both generally applicable and sufficiently detailed, especially as the GWP focuses on policies and strategies and not on IFRM projects. Moreover, few examples of IFRM projects have been identified, documented and evaluated (Pahl-Wostl et al., 2007) as the implementation of IFRM is still in its infancy (Huntjens et al., 2011) and evaluation frameworks for IFRM are not readily available.

This research aimed to deduce attributes of effective investment projects in developing and implementing IFRM based on a case study, and to validate these attributes by testing with literature on IFRM. Empirical evidence is analysed as to how IFRM came about (output and process) and what enabled and constrained the delivery of IFRM. Room for the River (RftR) is a large scale investment 2.3 billion Euro project that is nearly complete to deliver a programme for flood risk management. It has been selected here as a case study for several reasons. RftR is an exemplary project for IFRM that delivers integrated outputs (Rijke et al., 2012b) in terms of objectives, spatial scales and temporal scales (Section 3.1). RftR embraces an integrated approach by engaging multiple stakeholders and disciplines in the planning process (Van Herk et al., 2012; Section 3.2). As opposed to other case studies that comprise research or policy pilots or projects in initiation phases (e.g. Hegger et al., 2012), the RftR case study allows for the analysis of the integrated process in several planning stages. Many authors call for the application of IFRM, for practical guidance and evaluation. This paper presents empirical evidence for attributes of effective IFRM projects and programmes that can be used to design and evaluate future projects.

4.2 Research approach

A mixed-method case study approach has been used to devise attributes for projects to deliver IFRM. The results are confronted with literature in IFRM for a preliminary validation, to be compared with other case studies in future research. The research has been conducted over some 2 years in sequential stages by six researchers.

The integrated outputs of RftR (Section 3.1) have been analysed using document analysis of: programme documents (Ministerie V&W et al, 2006); spatial plans for the various projects; progress reports; and external audits. Semi-structured interviews (n=55) were conducted to understand how integrated output was brought about from the initial planning stages and sustained towards the implementation (integrated process). The interviewers asked specifically about the activities that had been conducted and the knowledge that was generated for use in subsequent planning stages (Section 3.2). 10 of the respondents were involved with the Initiation phase of the programme and 31 were directly involved in the Design and Realisation stages of the programme. Interviewees represented a range of disciplines and organisations involved with the individual projects (i.e. waterboards, provinces, municipalities and Rijkswaterstaat, the executive arm of the Ministry of Infrastructure and the Environment that is responsible for the design, construction, management and maintenance of the main infrastructure facilities in the Netherlands) and the programme as a whole (i.e. the RftR programme Directorate, its mother organisation Rijkswaterstaat and the Ministry of Infrastructure and Environment). 14 respondents were indirectly involved with or observers of the programme: national policy makers and senior policy advisors in water management and infrastructure; and managers of other IFRM programmes. Following the initial interviews, 13 more were conducted with project leaders, consultants and operators of 3 individual RftR projects that are in the Realisation phase: Noordwaard; Munnikenland; and Veessen-Wapenveld, to elicit how they sustained integration from the design phase towards implementation and subsequent operation and maintenance (Albers, 2012). Interview transcripts were analysed using QSR Nvivo 9. Data have been coded inductively for attributes of RftR that enabled or constrained integrated processes and outputs that emerged from the data. After additional literature review in IFRM and cross-sectoral collaboration (Bryson et al., 2006) the coding was revisited, definitions of attributes found in RftR were formulated in IFRM terminology, and evidence was classified. The evidence for the attributes from RftR were confronted with literature in IFRM for initial validation (Section 4).

After a year of research and the first round of interviews, a survey was conducted of the RftR participants. There were 151 survey respondents: 48 from the RftR programme office; 10 from other parts of Rijkswaterstaat; 10 from the Government Ministries involved; 11 from Provinces; 22 from waterboards; 36 from municipalities; 7 from the private sector; and 7 other respondents, such as scientists and community groups. The attributes that had already emerged from the data were incorporated into a set of questions to quantify: their importance for the success of RftR; their uptake beyond RftR; their correlation with stakeholder satisfaction about the programme and individual projects; and the contribution of stakeholders to integrated processes and outputs. Respondents were asked to rate the attributes, their satisfaction, or contribution using a five-point Likert-based scale, where 1 was very unimportant, very unsatisfied, or very small contribution; and 5 very important, very satisfied, or very large contribution. The survey responses were analysed statistically using IBM SPSS 19.0. Section 4 presents results for average scores and standard deviations for the attributes. E.g. the satisfaction of respondents with the results of the programme was 3.96 on the five-point scale, thus close to 'satisfied' (4), with a standard deviation of 0.69, thus tending down toward 'neutral' (3) and up toward 'very satisfied' (5). Correlations were analysed using the Gamma coefficient*. A confidence interval of 99% was selected and thus Section 4 presents only the Gamma coefficients for when the significance level of the correlation is 0.01 or more. The data were analysed for classifications per type of organisation (e.g. national government or municipality), or type of position (e.g. politician or professional). To validate the findings with practitioners, a network event discussed the lessons learnt from RftR with 150 participants that have been involved in RftR by asking for supporting or possibly contradicting evidence that are presented in Section 4.

* The gamma coefficient is calculated by counting the number of concordant pairs of cases in a contingency table, subtracting from this the number of discordant pairs and then dividing the result by the total number of pairs. The correlation is +1 in the case of a perfect positive (increasing) linear relationship (correlation) and −1 in the case of a perfect decreasing (negative) linear relationship (anti-correlation). The Gamma coefficient is used, as opposed to other methods such as Pearson correlation, because the data is ordinal and not equidistant from one another (i.e. the distance between 'very unimportant' and 'unimportant' is not necessarily the same as from 'unimportant' to 'neutral').

4.3 IFRM; case study Room for the River

Rijke et al., (2012b) show that the Room for the River programme has an exceptionally high performance in terms of project output, stakeholder satisfaction, budget and time when compared with other large water programmes in the Netherlands (Taskforce HWBP, 2012), or with many large international infrastructure projects (Flyvjberg, 2007). The flood safety objective of increasing the river discharge capacity for riverine areas of the Rivers Rhine, Meuse, Waal, IJssel and Lek from 15.000m3/s to 16.000m^3/s is being met, according to an independent evaluation of Deltares (PDR, 2011). The second objective of contributing to the improvement of the spatial quality of the riverine area is also being met (Hulsker et al. , 2011). In terms of budget, the Programme Directorate reported to Dutch Parliament that the total cost estimate for the programme was 2170.9 million Euro in 2011 compared with a budget of 2180.8 million Euro originally planned for in 2006 (PDR, 2011) As for the time planning, this progress report states that out of the 39 initial projects that were described in the policy decision in 2006, only 8 are expected to have a delay of approximately one year (completion originally scheduled for 2015). The results of the survey indicate that the actors involved are satisfied with the results of both the programme and the individual projects so far delivered, with 85% of survey respondents indicating that they were satisfied or very satisfied.

4.3.1 RftR integrated outputs

The question addressed here is if RftR has also delivered integrated output: in objectives, and across spatial and temporal scales. *Integration in objectives* is a direct consequence of the Programme's two objectives of flood safety and spatial quality and of the concept of river widening. Hulsker et al. (2011) show how functions such as: agriculture, recreation, nature, cultural-historic values and housing are integrated into flood safety projects. For example in Nijmegen and Deventer, the projects have provided opportunities for urban development through better connection of both sides of the river. In Overdiepse Polder and Noordwaard, agricultural land was preserved in inundation polders or depoldered areas. The concept of river widening comprises measures, such as flood by-passes, excavation of flood plains, and dike relocation (Fig. 4.1), that have a stronger spatial component than traditional measures such as dyke reinforcement. Put differently, river widening is a spatial measure by definition and hence requires integration in land use planning.

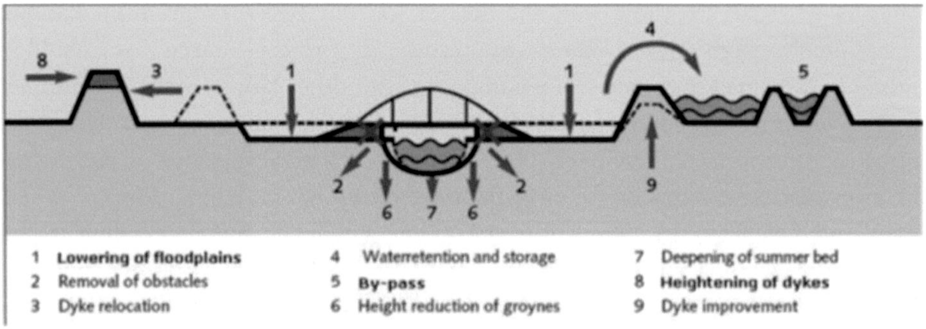

Figure 4.1. 'Measures that are applied in Room for the River (Source: Room for the River Programme Office)'

From a hydraulic perspective, all measures in the programme are related because their effectiveness to increase discharge capacity are mutually dependent. They cover the Dutch rivers Rhine, Meuse, Waal, IJssel and Lek from the German border on the East to the IJssel Lake in the North and the North Sea in the West. *Integration in spatial scales* has been commonplace during the Initiation phase when a collection of measures have been selected along the length of the river system. Also, this integration is apparent, as during the later planning phases, 5 of the 39 projects were cancelled as being superfluous, because more water level reduction had been achieved than originally planned at several other locations. Also, in individual projects, such as the Veessen-Wapenveld river bypass project, solutions were sought outside the original project area to overcome hurdles in the planning process. For example, agricultural land that had to be sacrificed in a flood prone area was traded for a nature area outside the project area to be converted in agricultural land. However, this trade-off occurred only rarely. With regard to spatial quality, the measures along a river branch could have been designed more coherently to contribute to a uniform Dutch river landscape. New civil structures and by-passes have not all been designed consistently, but rather by regional project teams for each individual project (Hulsker et al., 2011).

In its essence RftR comprises *integration across temporal scales*. Increasing the river discharge capacity helps to prepare for higher peak discharges now and in the future. RftR set an example for countries upstream of the Netherlands. It supported international agreement on EU Water Guideline legislation, stipulating that these neighbouring countries could not take measures that pass increasing water flows downstream towards the Netherlands. Also, river widening inherently provides future flexibility to implement complementary measures because future dyke reinforcement or heightening measures can still be considered after the river discharge capacity has already been increased. During the initiation phase of the programme, an evaluation was carried out to assess if RftR could accommodate the passage of 18.000m³/s in the river systems in the future instead of the 16.000m³/s originally planned for. It was analysed if measures could retain their functionality and have a 'no-regret' performance, being useful under any future scenario, such as for a discharge of 18.000m³/s (Schut et al, 2010; Van Herk et al., 2012a).

4.3.2 RftR integrated process

How did the planning process deliver integrated outputs? Van Herk et al (2011a) provide a description framework that distinguishes three types of activities in IFRM processes: system analysis; planning, design and engineering; and governance. Here we distinguish 4 planning phases as shown in Figure 4.2. Table 4.1 summarises the activities per planning phase.

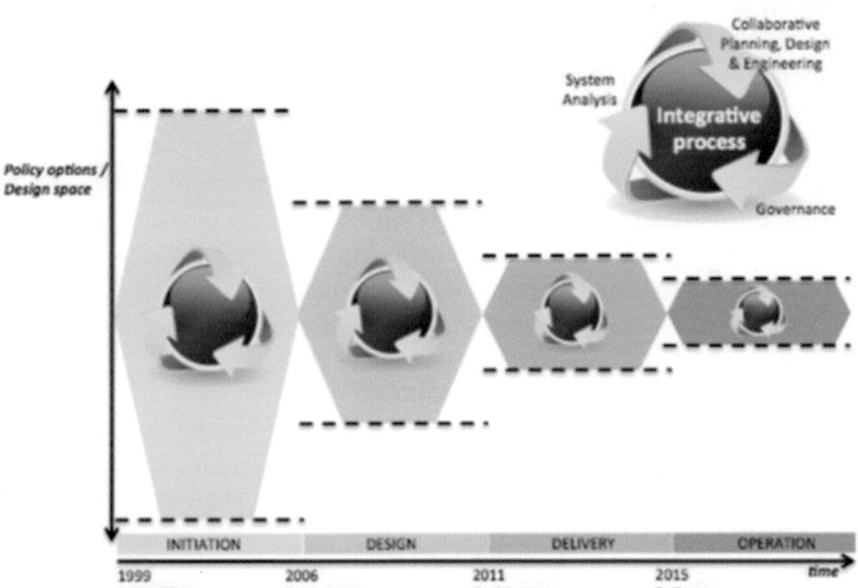

Figure 4.2. 'The integrated process through the planning phases with decreasing design freedom`

The Initiation phase for RftR started in 1999 with the national policy on Room for the River that merely prescribed the objective to create a river discharge capacity of 16.000m³/s by river widening measures. Solutions could be devised for the entire river system and from a long list of 700 possible measures, 39 were selected and approved by parliament in the Programme Decision Room for the River in 2006 (V&W, 2006). From the outset the *design* activites made a crucial contribution to integration by the choice of river widening as a solution and thus as the leading design concept for all measures. The concepts of river widening took shape in the individual measures proposed that combined multiple objectives. In the *governance* activities the Dutch Government set the objectives of both flood safety and spatial quality. The spatial quality objective was advocated by the Ministries of Housing, Spatial Planning & the Environment and of Agriculture, Nature and Fishery for whom spatial quality is their core interest in RftR. They jointly commissioned RftR with the Ministry of Public Works and Water Management who are responsible for flood safety. A related *governance* decision has been to provide regional governments design freedom to propose and design measures and to grant them leadership to implement these individual measures. Regional authorities themselves proposed the 700 possible measures, because there was something 'in it for them'. The collaborative process is exemplified by the computerized hydraulic model/scenario planning tool called 'box of blocks' (*blokkendoos* in Dutch). The tool embodies *system analysis*, but also connected with *design* and *governance* activities for the entire river system. Schut et al. (2010) explain how the instrument evolved from first a tool used by hydraulic engineers (also Reuber et al, 2005) to 'explore solution space' and calculate the hydraulic consequences of a combination of river widening measures, to later supporting the design and selection of measures, facilitating dialogue, cooperation and eventually decision-making between politicians from different levels and regions. The tool was made available to all stakeholders to 'play with' and could demonstrate and visualise the effectiveness and interdependencies of measures to increase discharge capacity and thus reduce water levels.

The design freedom during the Design phase was limited to the 'scope' of each of the 39 measures as prescribed in the Programme Decision. It stipulated per measure: the type of measure such as lowering of a flood plain or creating a river by-pass (see Figure 4.1 for all types of measures); water level decrease aimed for; the national budget assigned; and the Initiatior, or leader, of the planning process. The designs of these measures could be developed freely by the Initiator and functions could be coupled, especially to enhance spatial quality. E.g. the design of the river by-pass at Lent included waterfront developments with recreational functions as proposed by the municipality and inhabitants, and nature development as proposed by the regional government. The *collaborative design* processes varied per project. The project team organised design sessions with the participation of various stakeholder groups, including the local inhabitants. The

Ministry periodically brought in an independent team of experts named the Quality-team, to suggest improvements to the plans and to inspire local project teams and decision makers. Following that example, local Quality-teams were installed and the project teams have been assisted by landscape architects. Design work was supported by *system analysis*. The Programme Directorate prescribed mandatory analysis of e.g.: soil quality, archaeology, cabling, compliance with local and regional land use planning. These analyses were required to avoid surprises later in the process such as encountering electricity cables or archaeological artefacts where soil is to be removed that could then delay the project. *Governance* activities mainly focused on the politics of discussing ambitions and budgets for measures. The project team proposed these to the steering committees whose members brought them for formal decision making to their respective democratically elected parliaments. Such discussions were especially valuable as for some projects the regional governments provided up to 50% of funding to combine and integrate regional and local policy objectives, with the flood safety objective that was funded by the national government.

The realisation phase started during 2011 and beginning of 2012, depending on the progress on each project. By that time the design freedom had decreased considerably and was limited to the Project Decision. Contractors were, however, allowed some design freedom to select materials and optimize the construction process. Integrated outputs had been anchored in the tender contracts based on the final project design for each individual scheme. Politicians from the steering committees and civil servants from the project teams tried to streamline building permit requests across the various authorities. An example for one authority was that they would ideally have wanted construction planned at night to reduce traffic nuisance, whereas for ecological reasons work at night was not recommended. An important test for assessing the effectiveness of the IFRM process of RftR will be in the future operation and maintenance phase. Many interviewees indicated concerns about the lack of agreement on budgets and funding for operation and future quality controls of functions. E.g. the Munnikenland project has suffered delays in design and construction due to discussions about operational tasks and budgets for future maintenance. The municipality will own more hectares of natural land increasing the burden of maintenance, whilst in the national parliament there are talks of budget cuts for nature operation and decreasing funding for the State Forestry Agency that could support the municipality.

Table 4.1. Process integration applied for RftR

Activities	Initiation	Design	Realisation*	Operation*
System analysis	Analyse river discharge capacity; and cost & effects of measures	Assess hydraulic performance of measures. Analyse technical feasibility; financial feasibility; risk analysis; planning & costs.	Analyse the construction process.	Monitor performance of functions & values
Collaborative planning, design & engineering	Choose river widening concept. Propose 700 measures.	Design alternative measures. Propose improvements by Quality-team and architects.	Select materials and devise construction process.	Define operation plans partly depending on tasks & budgets
Governance	Set two objectives: flood safety and spatial quality. Select 39 measures and provide mandate for related Programme decision.	Discuss and decide on ambitions and budgets for measures.	Organise public-public-private tender processes and define contracts that embed integration. Coordination permit requests.	Decide on tasks & budgets (yet to be done)

*as planned, not observed

4.4 Attributes for IFRM projects

From the interview data two types of attributes for IFRM projects emerged: those that directly foster IFRM processes and outputs and those that enable the implementation of IFRM projects. These attributes of the former type (Table 4.2) were identified during the analysis of integrated outputs (Section 3.1) and processes (Section 3.2). Their wider applicability is discussed here based on an additional literature review related to IFRM. Attributes of the latter type, enabling implementation, do not directly stimulate integrated processes or outputs in themselves, but were mentioned by interviewees as being instrumental in implementing the RftR programme and its constituent projects. These have been compared with literature here to validate these enabling attributes for implementation,. Bryson et al. (2006) devised 22 propositions for cross-sector collaborations of the type needed for IFRM, based on literature from a variety of policy domains in the public sector. All of the 22 propositions have been found applicable to the RftR case study described here and have been related to the attributes to implement IFRM as deduced from the case study (Table 4.3). These attributes are being discussed in the following Sections 4.1 and 4.2 respectively.

4.4.1 Attributes to foster IFRM

An *integrated vision and concept* provides a visual connection to problems and potential solutions that inherently embraces integrated outputs and processes. Hence interventions can contribute to different policy objectives and therefore appeal to multiple stakeholders encouraging them to participate in the planning process. In RftR the integrated concept was 'river widening'. Section 3.1 explains how the concept inherently supported integration of objectives and crossed spatial and temporal scales. Three interviewees stated how they saw the integrated essence of the concept: ¨river widening is a spatial measure that requires (physical) integration by definition¨. Survey respondents scored their satisfaction with river widening on average as 3.97/5. River widening is positively correlated with multiple objectives (gamma=0.413), indicating that the concept enabled the delivery of multiple objectives. Several interviewees stated the importance of the 'brand name: Room for the River' that provided a visual connection to the problem and the integrated solution. Examples abound of flood management projects that are not based on an integrated vision. E.g. the US National Levee Safety Programme has an inherent sectoral vision to assess and upgrade levees (ASCE, 2011). That programme was first approved in 2006 after Hurricane Katrina devastated the city of New Orleans and is unlikely to deliver integrated outputs (Kates et al., 2006). Other integrated visions and concepts for IFRM have also been found. 'Building with nature' in the Netherlands integrates *inter alia,* coastal protection with land use planning through land reclamation using sand from the sea for dunes and beaches (Waterman, 2008). 'Flood proof building' promotes housing development in flood prone areas that can withstand flooding (Van Herk et al., 2011a). The concept of 'Green infrastructure' (Sandström, 2002) aims to utilise green systems as much as possible in urban space, that if done properly can help reduce and manage flood risk (Gill et al., 2007; Tzoulas et al., 2007). The disadvantage of an overarching vision can be that if it is too narrow it can stipulate the type of solution to be adopted and inherently reduces design freedom for a project. On a national level and for the long term, a strong vision could be the 'new' paradigm into which policies are 'locked-in', just as has happened with the previous paradigm: 'fighting against water'.

Setting *multiple objectives* for a project or programme from various policy domains makes integration of objectives an explicit aim. RftR had two objectives: flood safety and spatial quality. Adding the spatial quality objective appealed to many stakeholders that were interviewed: "the double objective [including spatial quality] enabled shared ownership". Respondents' satisfaction with spatial quality correlates to overall satisfaction (gamma 0.595) and flood safety (gamma 0.604). The latter indicates that the second objective supported delivery of the primary flood safety objective. The spatial quality objectives provided access to additional funding mechanisms. E.g. in the Deventer project the regional government co-funded a nature-farm in the outer marches. The double objective was an important success criterion that scored 4.03 out of 5 with a standard deviation of 0.66. Multiple objectives increased stakeholder satisfaction (positive correlation with significance of 99% and gamma of 0.652); increased general public support (gamma of 0.486); and according to project managers (n=17) it led to delivery within time and budget (gamma of 0.739). This attribute seems straightforward, but many water management projects lack multiple objectives because of the potentially increased political complexity for the planning processes (Woltjer and Al, 2007). If flood safety is not an explicit or the primary objective as in RftR, incorporating the flood risk consideration into urban or land use planning is hampered. E.g. the Thames sewer tunnel project in the UK aims to manage pollution in the river Thames, but despite its potential does not contribute to flood risk management (Brown et al., 2011). If flood risk is not set as an objective, considering flood risk along other objectives can be obliged through regulation or procedures, such as strategic environmental assessment (Therivel, 2012) or the 'water assessment' in the Netherlands (Neuvel and Van den Brink, 2009). Otherwise it can be incentivised through transfer of responsibilities and insurance premiums (Treby et al., 2006) or advocated from e.g. shadow networks (Olsson et al., 2006; Huitema & Meijerink, 2010).

Delimited design freedom provides the scope to devise integrated outputs within an integrated vision to deliver upon multiple objectives. During the Initiation phase of RftR, objectives were set at the level of the river basins within the Dutch borders, without prescribing the location of measures along more than 500km of river. At the start of the Design phase, objectives for individual measures were set in terms of reduced water levels for a river segment of several kilometres, leaving freedom to select the exact location and design of a channel excavation or dyke relocation. As one project leader from a municipality said: "RftR provided freedom for new solutions and enrichment of designs, especially to incorporate our local interests". The design freedom was, however, constrained within the vision of river widening and within the objective of reaching a river discharge capacity of 16.000 m3/s. The vision and objective were derivatives from an overarching flood safety objective and did not allow for investments from the programme budget in e.g. evacuation plans that can reduce potential casualties from flooding. Also, the river discharge capacity objective of 16.000m3/s implicitly enforced the existing legal design standard for levees to protect the Dutch river areas from a flooding event that is expected to occur once every 1250 years. This limitation to integration across temporal scales has been partly offset by scenario analyses that assessed the possibilities of taking additional measures in the future to cope with possible higher return periods (Section 3.1). In general, design freedom can be increased either by setting objectives and boundary conditions at the highest possible spatial scale level (e.g. river system instead of a single structure such as a bridge or levee) and with lowest degree of specification (objectives such as flood safety instead of derived functional or technical specifications) (Van Herk et al., 2006). A flood safety objective that explicitly aims to reduce the potential consequences of flood events (e.g. casualties, damage, societal disruption), as well as an objective to reduce the flood probability, leaves more design freedom than design standards for flood defence systems that are set based on return periods that are traditionally regulated in many countries (Jonkman, 2007). However, there are likely to be limits to increasing design freedom and there is no evidence available yet of an increased effectiveness in delivering flood safety by projects that set objectives with a lower degree of specification. For example, the Dutch Delta Programme is exploring the potential of stretching the design freedom further in research projects that formulate a flood risk objective in terms of probability and consequences and promote a 'multi layer safety' approach that considers a combination of measures comprising flood protection, spatial planning and emergency planning (V&W, 2008). Analysis of the projects in the Delta programme shows that the design and implementation of packages of measures are hampered by existing approaches, policies and regulations that are based on a paradigm of flood protection that, rather than IFRM: does not yet recalibrate a distribution of tasks and responsibilities; does not yet allow re-allocation of funding; and complicates the way in which these alternative packages of measures are understood by politicians and the general public (Kolen et al., 2010; Van Herk, 2013a).

The planning process should comprise *mutually enriching activities*. IFRM projects involve complex decision-making that requires knowledge generation and transfer (Van Buuren, 2006) that is stimulated when the activities: system analysis; planning design & engineering; and governance, are collectively mutually enriching (Van Herk et al., 2011a). Section 3.2 shows how these 3 activities have undergone a process of divergence (knowledge development in one activity) and convergence (knowledge exchange between activities) in each planning phase of RftR. The mutual enrichments of proposing (design), analysing and selecting (governance) contributed to the integrated nature of the outputs particularly during the Initiation phase. E.g. the Mayor of Gorinchem proposed river widening to be combined with the development of an industrial area in the outer marches at Avelingen (design). Analysis showed that the measures could reduce water levels at a low cost and that: ¨ the measures allowed our City's most important employer to expand its business there; an important reason for our support¨ (governance). In terms of integration in time scales, the *analysis* of the potential to deal with higher discharge capacity fostered the political dialogue on ambitions (*governance*) and spatial reservations that allow for possible future measures (*design*). Examples have also been found in RftR where activities did not enrich each other, nor supported an integrated process; but rather led to a sectoral approach. According to one interviewee: ¨the mandatory requirement to develop extreme alternatives, one being the 'cheapest' and the other being the 'nicest', helped in exploring the design freedom, but tended to separate the objectives of safety and spatial quality that became viewed as opposites in terms of options; whereas they can be combined and integrated¨. In general, there is no blueprint for the organisation of integrated processes (Van de Ven et al., 2011; Van Herk et al, 2011a). The attribute of mutually enriching activities is broadly defined here with the ambition to be applicable in other IFRM projects, but this requires further validation.

Collaborative planning & public participation allows individuals and organisations to enrich the integrated process with knowledge and means to develop and implement integrated outputs. Interviewees provided examples of stakeholders' contributing to integrated solutions for all RftR projects,. E.g. the alderman of Zaltbommel indicated that the municipality proposed a recreational hotspot with: a car parking lot; a quay for mooring boats; and an intersection of bicycle routes. In the Overdiep project, local formers designed by themselves a polder that is designated to be floodable with a return period of 1/100 years to reduce the water level in the Bergsche Maas River in order to protect the urban areas around Hertogenbosch. The agricultural function of the polder will be maintained and their farmhouses and businesses are being rebuilt on earthen mounds. 139 out of 152 survey respondents indicated that they perceived to have enriched the vision, ambitions or objectives of the programme or of the projects. On average they scored their contribution between 'significant influence' and 'high influence': 3.31/5. The enrichment indicated by respondents that worked on regional projects was found positively correlated with their general satisfaction (gamma=0.419) and satisfaction with the way the? integrated measure was designed (gamma=0.417). For politicians that correlation is higher at gamma=0.632. Hence, it can be concluded that in RftR, collaborative planning and public participation has contributed to its 'perceived' effective delivery of integrated outputs. Collaborative planning has been enabled by regional freedom (Section 3.2) and respondents of the survey scored 'regional freedom' as an 'important' success factor for the delivery of RftR: 3.9/5. It is however questionable whether collaborative planning and public participation always support (perceived) effective delivery of integrated outputs. The integrated process in RftR had an institutional fit because enhancing flood safety fits in the unitary dimension of the Dutch institutional structure. The spatial quality objective activated the decentralization dimension as lower level government have the right of initiative and discretionary power as long as it is not inconsistent with policies formulated at a higher level (Toonen, 1990). The result was that municipalities, provinces and water boards started embracing the programme to couple the river widening measures to their own ambitions. An integrated process inherently requires collaboration between stakeholders that have various interests at various spatial scale levels and for various temporal scales and contribute various disciplinary knowledge and means. Edelenbos et al. (2011) show how stakeholders can contribute their 'lay knowledge' through public participation to enrich solutions and increase legitimacy of decisions. However, there is no blueprint for the organisation of collaborative planning and public participation. Huntjens et al. (2011) and Zevenbergen et al. (2013a) conclude from international comparisons of IFRM projects that the organisation and effectiveness of collaborative planning and public participation depends on the context, e.g.: the institutional; cultural & socio-economic; historical context; and geographical contexts. Table 4.2 summarises the attributes that foster IFRM.

Table 4.2. Attributes to foster IFRM

Attributes to foster IFRM	Definition and applicability to IFRM
Integrated vision & concept	An *integrated vision and concept* of a project or programme provides a visual connection to problems and potential solutions that inherently embraces IFRM, such as: river widening; building with nature; flood proof building and planning; green infrastructure.
Multiple objectives	The formal objectives of a project or programme comprise various policy domains such as flood safety; housing; recreation; nature; mobility, and hence engage different stakeholders and funding mechanisms.

Delimited design freedom	The design freedom to devise integrated outputs within the boundaries that are set at the highest possible scale level and with lowest degree of specification. E.g. flood safety in terms of casualties and expected damage, instead of technical design standards for levees or drainage systems for specific return periods. Or river discharge capacity (m3/s) that instead of prescribing, leaves freedom to select the location and design a type of measure.
Mutually enriching activities	The activities: system analysis; planning design & engineering; and governance, can be mutually enriching and undergo a process of divergence (knowledge development) and convergence (knowledge exchange) within progressively narrowing boundaries (design freedom) in subsequent planning phases as the project advances (Figure 4.2).
Collaborative planning & public participation	Multiple individuals and organisations enrich the integrated process with knowledge and means to develop and implement integrated outputs, rather than a process that only involves a limited number of stakeholders with legal mandates.

4.4.2 Attributes to implement iFRM

Exploit disasters and failures of sectoral approaches. The survey respondents ranked the 1993 and 1995 near floods as the most 'important' success factor (4.3/5) for the implementation of RftR compared with organisational, human and contextual factors that scored from 3.2 to 4.2 out of 5. According to interviewees, the near flood events gave a sense of urgency to deliver the programme, but that does not explain the choice for an integrated concept or multiple objectives. Two of the local politicians interviewed that were involved from the start of the RftR process listed a combination of factors that they believed had promoted an integrated process rather than a sectoral approach for RftR. These were: the deputy minister wanted a paradigm shift from 'fighting against water' to 'living with water'; initial pilot projects such as Plan Ooievaar had already experimented with agriculture and nature development in flood plains; and local resistance to dyke reinforcement projects, the perceived alternative to river widening, was fierce in the 1990s. Additionally, Rijkswaterstaat employees indicated the importance of the lessons from the poorly delivered High Speed Rail and Betuwe rail freight route projects. These lessons were used to help create regional public support to stimulate project progress via an integrated process, as opposed to the local resistance the earlier projects' sectoral approach had provoked. After a flood disaster large investments are made for recovery and redevelopment and to reduce flood risk for the future (Zevenbergen et al., 2013b). Many projects after disasters aim to reduce flood probability through investments in flood defence infrastructure based on politicians' misleading ambition 'to prevent a disaster from *ever* happening again'. This happened previously in the Netherlands, where the Delta Works were initiated as a response to the 1953 floods (Zevenbergen et al, 2013b). After hurricane Katrina damaged New Orleans, the US Army Corps of Engineers invested billions of US dollars in the flood defence infrastructure, thus continuing 'rebuilding the familiar' and risking the 'levee-effect' (Kates et al., 2006) with further technical lock-in to outmoded structural solutions (Walker, 2000). In contrast, Room for the River provides an example showing that disasters can create an opportunity to abandon a sectoral approach. Shifting the focus from flood protection to reducing the exposure and vulnerability to floods requires by default, integration of flood risk management with spatial planning (Yovel, 2013).

Implementation of an integrated project requires *committed sponsors that assign resources through formal agreements.* Interviewees indicated the importance they attached to the 'Programme Decision' formal agreement having adopted the river widening concept and double objective from the outset, and anchored in the assigned national budget. The Programme Decision was promoted by 3 different Ministries and unanimously adopted by Dutch Parliament. Regional government representatives indicated that this triggered them to commit additional budgets because national funding had been secured. Integration was further anchored in transfer documents between planning phases and stakeholders (Albers, 2012) right up to the contractors realising the projects. E.g. spatial quality was a selection criterion in tender procedures and was detailed in accompanying ambition documents. Also, interviewees indicated the importance of balancing progress in the planning process with quality and budget controlling based on the Programme Decision. The Programme Directorate introduced a justification cycle that comprised checks on multiple facets of feasibility (financial, technical, etc) for milestones in the decision-making procedures. According to the survey results, clear objectives and the tight steering for milestones have been 'important' (3.99/5) to successfully implement RftR. Moreover, the success factor of clear objectives and tight steering is positively correlated with the satisfaction about increased water safety (gamma=0.43) and delivery within budget and time constraints (gamma=0.41). A formally assigned task to deliver the project leads to a focus on technical, legal and financial feasibility of implementation. Many research and policy initiatives on IFRM that have been carried out by informal networks (e.g. Farrelly and Brown, 2011) were often in an initiation or policy definition phase and had not (yet) been confronted with implementation and feasibility demands, as in contrast happened in RftR that has been through multiple planning phases. Formal agreements are to be sustained or be renewed after each planning phase that narrows the boundaries on scope. A funnelled planning process reduces design freedom with milestones that are anchored in decision-making procedures that simulate progress and ensure integration towards implementation (Edelenbos et al. 2011).

The integrated *process is to establish legitimacy, build trust and embed accountability systems* to safeguard support for integrated outputs and for the integrated process itself. In RftR, the legitimacy of the integrated process and outputs was obtained through: the unanimous parliamentary support for the Programme Decision with its integrated vision and double objective; the collaborative inventory and selection of integrated measures; the regional leadership in design, construction and operation of measures; all combined with regional public participation processes. Legitimacy has been increased further by audits and evaluations by independent scientists (e.g. Twist et al, 2011) and the national planning bureau that conducted an independent cost-benefit analysis about RftR prior to Government approval (Ebregt et al., 2005). For the purpose of accountability in RftR, formal independent audits have been performed periodically during the programme on both objectives: river discharge capacity by Deltares, a research institute (PDR, 2011), and spatial quality by Ecorys, a consultancy firm (Hulsker et al. , 2011). Furthermore the justification cycle operationalized accountability and required collaboration between all stakeholders, from regional projects to the Programme Directorate and to Dutch Parliament (Rijke et al., 2012b).

Interviewees highlighted that trust was enhanced through this continuous collaboration, as well as through the apparent legitimacy and accountability. The survey results indicate 'satisfaction' with RftR outputs (4.1/5 overall) that is positively correlated with the satisfaction with collaboration (gamma=0.71) and public support (gamma=0.81). However, the survey data also show a few outliers of local respondents that indicated dissatisfaction. These have been explained from interview data that show how trust decreased in some projects. In the Veessen-Wapenveld project, local farmers stepped out of the participatory planning process after their suggestions were disregarded as 'suddenly' being out of scope. According to project team members the project scope was not clearly communicated upfront jeopardizing the trust of participants. In the Noordwaard project, trust of local inhabitants suffered when the Deputy Minister's promise of 'generous compensation for expropriation' was not followed up for several years by a definite proposal. Transparency in decision-making has been mentioned by interviewees and later during the network event for validation, as a recommendation for future projects to build trust, whilst also discerning that too much transparency could reduce freedom for political manoeuvring to agree on integrated outputs. Network management strategies can support legitimacy, trust and accountability, and vice versa, but there is not a single blueprint to devise network management strategies (Agranoff and McGuire, 2001), nor for integrated processes (Section 1). Several researchers provide general recommendations in line with the findings from RftR. On legitimacy, Cuppen (2007) concludes that enhancing legitimacy requires the creation of broad support that in turn helps the implementation of IFRM. Edelenbos & Klijn (2006) state that interactive decision-making raises the legitimacy of decisions. Trust in governance networks is important for achieving better (perceived) outputs and can be developed and sustained by network management (Klijn et al, 2010). Accountability systems track the implementation of integrated outputs and are built on strong relationships with key political and professional constituencies (Bryson et al., 2006).

A collaboration *structure is to be fit-for-purpose and adaptable: balance top-down and bottom-up governance; and formal and informal relations.* In RftR the national government set the boundary conditions for the integrated processes and outputs (objectives, scope, constraints), but endorsed regional governments with leadership over planning, construction and operation of measures (Section 3.2). This multi-level governance arrangement was mentioned by all interviewees involved in RftR's initiation phase as a success factor and scored in the survey as an 'important' success factor for the implementation of the Programme (3.9/5). The main relations between national and regional authorities had been formalised and guided the integrated process, but many informal relations also influenced the integrated process. Informal and ad-hoc collaboration will always take place in planning processes (Hillier, 2000), but in RftR specially assigned stakeholder managers have monitored informal interactions as well as the interests of various stakeholder groups. Moreover, programme and project managers have tried to embed informal networks into the planning process through the creation of advisory boards in which lobby groups or advocacy coalitions could participate. E.g. the farmers that opposed the river by-pass to be excavated in Veessen-Wapenveld through their farmland, lobbied the Deputy Minister and MPs (of a pro-agriculture political party) directly and via the farmers association. The Programme Directorate engaged in discussions with the politicians and farmers, invited them back into an advisory group and conducted a study on how to preserve or create economically feasible farmland for them and for other farmers affected by RftR. Based on this variety of lessons, it is difficult to prescribe collaboration structures or their attributes to implement IFRM projects more concretely. RftR had the Programme Directorate as lead-organisation, but this single case study research cannot provide evidence that such a structure would have been more effective than other network governance structures such as (Provan and Kenis, 2008): shared governance networks; or network administration organisation (NAO) governed networks. Moreover the structures in RftR have been flexible and have been adapted when opportune, as summarised by the Programme Director: ¨structure should follow strategy¨. The overarching concept that guided the structure of collaboration between national and regional governments was stable, but elements of that structure have been adjusted several times during the process, such as meeting, reporting and communication procedures, and department or project team structures. Many other authors use a descriptive and analytic, rather than prescriptive perspective to collaboration structures. Huitema et al. (2009) discuss the features of polycentric governance in IFMR in contrast to a classical modernist approach to institutional design that Hajer (2003) considered unfeasible, ineffective and inefficient. Polycentric governance, with multiple centres of power, has several advantages: it is more resilient and can better cope with change and uncertainty (integration across temporal scales); can manage issues with various geographical scopes at different scales (integration across spatial scales); and the different organisations can learn from each other. Disadvantages include the potential loss of democratic accountability; the difficulties related to collective decision making; and the required coordination that has transaction costs. Huntjens et al. (2011) concluded that water management requires balancing between bottom-up governance (decentralisation) with top-down governance (centralisation) for e.g.: facilitation, capacity building, and cooperation across boundaries. Programmes, as opposed to projects or generic policies, can be structured and managed as hybrids that embrace polycentric governance within a classical institutional context and coordinate top-down and bottom-up governance (Rijke et al., under review).

Learning and feedback loops to *absorb threats and seize opportunities from a dynamic context* to implement integrated processes and outputs. RftR has been influenced by the significantly changed political landscape after the 2010 national and local elections. Political support for RftR's spatial quality objective decreased and in general government budgets for nature development and recreational development were cut, whilst agriculture as economic activity gained political relevance. At programme level and for individual projects solutions were sought to preserve or develop farmland. The communication director of the Programme gave instructions to: "play down the relevance of spatial quality in communication about the Programme to leave this element under the political radar". The implementation of European legislation on nature preservation (Natura2000) posed a threat as it limited the design space by fixing nature areas that could not be altered. The Programme Directorate and regional stakeholders sought creative solutions and lobbied to the European Commission to preserve and develop nature in adjacent areas, if nature preservation was challenged in the project area by the river widening measure. The changing context also presented opportunities. The new Water Act that came into force regulated the applicability of different soil qualities and replaced the Surface Water and Contamination Act and Soil Quality Licence. This enabled the realisation of different spatial functions with different soil qualities. RftR set precedents for the implementation of this new regulation and used it to increase collaboration between authorities on soil permits. Not only was the dynamic context in need of learning, but also RftR required much learning that has been instrumental in implementing integrated processes and outputs within an institutional and policy context that was designed and has evolved based on a sectoral approach (Van Herk et al., 2013b; in press). Many types of learning have been observed related to new concepts that RftR embodied such as: river widening; collaborative arrangements between national government and regional stakeholders; participatory processes regionally (Van Herk et al. (2013b; 2013c; in press). Rijke et al. (in press) conclude that the effectiveness of RftR programme management had been increased by its learning potential. In general, contextual factors can provoke time-delay; cost overruns and changes of scope in the implementation of large infrastructural projects and require learning (Hertogh et al., 2008). Feedback loops are necessary to be flexible and adaptable for many socio-economic and biophysical system changes that can occur during the implementation of programmes with a long duration, or of adaptation pathways that can have time horizons up to e.g. 50 years or more to reach a flood risk objective by combining multiple investment projects (Gersonius, 2012). For the same reasons these feedback loops are instrumental for the implementation of adaptive delta management (Zevenbergen et al., 2013b). Much recent literature presents research into learning for integrated and adaptive water management (e.g. Ashley et al, 2012; Mostert et al., 2007; Huntjens et al., 2011; Pahl-Wostl et al., 2012). Table 4.3 summarises the attributes to implement IFRM related to propositions on cross sector collaborations.

Table 4.3. Attributes to implement IFRM related to propositions on cross sector collaborations (CSC) (Bryson et al., 2006)

Attributes to implement IFRM	**Related propositions on cross sector collaborations (CSC) (Bryson et al., 2006)**
Exploit disasters and failures of sectoral approaches	P1. CSCs are more likely to form in turbulent environments. P2. CSCs are tried when separate efforts of different sectors have failed or are likely to fail.
Committed sponsors that assign resources through formal agreements.	P3. CSCs are more likely to succeed when linking mechanisms, eg powerful sponsors, or existing networks are in place. P4. The form and content of initial agreements and processes to formulate them affect the outcomes of the collaborations. P5. CSCs are more likely to succeed when they have committed sponsors and effective champions at many levels.
The planning process is to establish legitimacy, build trust and embed accountability systems	P6. CSCs are more likely to succeed when they establish legitimacy. P7. CSCs are more likely to succeed when trust-building activities (such as nurturing cross-sectoral and cross cultural understanding) are continuous. P10. CSCs are more likely to succeed when their planning makes use of stakeholder analysis. P20. CSCs are most likely to create public value when they are resilient and engage in reassessments P21. CSCs are more likely to be successful when they have an accountability system

Structure is to be fit-for-purpose and adaptable: balance top-down and bottom-up governance; and formal and informal relations.	P8. CSCs are more likely to succeed when partners use resources and tactics to equalize power and manage conflict effectively. P9. CSCs are more likely to succeed when they combine deliberate and emergent planning. P13. Collaboration structure and the nature of the tasks performed at the client level are likely to influence a collaboration's overall effectiveness. P14. Formal and informal governing mechanisms are likely to influence collaboration effectiveness. P15. Collaborations involving system-level planning activities are likely to involve the most negotiation, followed by collaborations focuses on administrative-level partnerships P17. Competing institutional logics are likely within cross-sector collaboration and may influence agreement on process, structure governance and desired outcomes. P18. CSCs are most likely to create public value when they build on individuals' and organisations' self-interests and strengths and compensate for weaknesses
Learning & feedback loops to absorb threats and seize opportunities from a context that is dynamic due to e.g.: political and policy change; economic cycles.	P11. Collaborative structure is influenced by environmental factors such as system stability and the collaboration's strategic purpose. P12. Collaborative structure is likely to change over time because of ambiguity of membership and complexity in local environment P16. CSCs are more likely to succeed when they build in resources and tactics for dealing with power imbalances and shocks. P19. CSCs are most likely to create public value when they produce first-, second- and third-order effects.

4.5 Conclusions

Most of the literature used for this research (e.g. Huntjens et al., 2011; Pahl-Wostl et al., 2012) calls for IFRM and, often implicitly, takes a normative stance to IFRM following organisations such as the Global Water Partnership, although Jeffrey and Gearey (2006) argue that empirical evidence is missing that unambiguously demonstrates the benefits of these integrated approaches. The research described here circumvents this debate by using Room for the River (RftR) as a case study that embodies IFRM and has also delivered on its objectives, within time and budget constraints and obtained high stakeholder satisfaction (Section 3). The relevance of the research is illustrated by many new projects that, despite clear guidance, are now starting to aim for integrated processes and outputs (Butterworth et al., 2010). Projects in particular, in addition to policy reform and institution building, have the potential to deliver IFRM (*ibid*). Following the definition of IFRM (Section 1) these projects will require: an approach to develop and implement measures to reduce flood risk by collaboration between multiple disciplines; a group of stakeholders with various interests and means; a combination of objectives and funding from different policy domains; consideration of a range of possible options at all spatial scale levels and for various time horizons. Literature provides limited guidance as to how to organise for IFRM (Section 1). The research described here has devised attributes of effective investment projects in developing and implementing IFRM based on a multi-method case study analysis, and these have been considered in relation to literature. It has been shown how RftR has delivered integrated outputs: in objectives; across spatial scales and time scales; through an integrated process that combined: system analysis; planning, design and engineering; and governance activities.

From the interview data two types of attributes for IFRM projects emerged: those that directly foster IFRM processes and outputs and those that enable the implementation of IFRM projects. The attributes of the former type are: to start with an *integrative vision and concept;* set *multiple objectives*; provide *delimited design freedom*; organise for *mutually enriching activities*; and for *collaborative planning & public participation*. This research provides evidence that each of these attributes has stimulated IFRM, but also that they have been applied simultaneously and are mutually reinforcing. It is clear that these attributes have been embraced from the outset, during the Initiation phase of the RftR programme. Attributes of the latter type, enabling implementation, do not directly stimulate integrated processes or outputs in themselves, but have been instrumental in implementing the RftR programme and its constituent projects. Attributes that enable the implementation of IFRM are deemed necessary as many water management projects start with the ambition to integrate, but these ambitions often do not materialise or projects do not get implemented at all (Edelenbos and Teisman, 2013). To support implementation of IFRM projects, *disasters and failures of sectoral approaches are to be exploited* to adopt an integrated concept and approach and set multiple objectives. Integration outputs are to be anchored through *committed sponsors that assign resources through formal agreements*. The collaborative planning *process is to establish legitimacy, build trust and embed accountability systems* to safeguard support for the integrated process and outputs throughout the planning process. The collaboration *structure is to be fit-for-purpose and adaptable: balance top-down and bottom-up governance; and formal and informal relations*. The planning process needs to stimulate *learning and feedback loops* to *absorb threats and seize opportunities from a dynamic*, especially as IFRM is still in its infancy. The evidence stresses that these attributes are intrinsically related to the previous 5 attributes that foster IFRM. E.g. an integrated concept such as river widening, requires spatial integration, this demands collaborative planning and participatory planning that in turn requires fit-for-purpose collaboration structures and processes. The attributes are hypothesised to be widely applicable to IFRM based on the literature review (Section 4). However, they need be customised to the environmental problem targeted by each project and to be aligned with the existing institutional and governance systems. Transition management researchers have studied the institutional barriers for IFRM (e.g. Van der Brugge and Rotmans, 2007), how these can be overcome (De Haan and Rotmans (2011), and how individual projects can contribute to this (*ibid*). Biswas (2004) doubts whether institutional and organisational integration for IFRM is possible and instead recommends coordinated collaboration between existing institutions, something that IFRM projects such as RftR have to do anyway. This research has only considered one in-depth case study and cannot therefore compare the results with additional cases. Comparative studies between a number of IFRM projects could possibly reveal further insights into the enabling and constraining factors for IFRM. The attributes that have emerged from this research can help inform future (case study) research and can be subsequently enriched by this. The definitions and clustering of attributes should be critically reviewed based on multiple case studies, because a single empirical test cannot validate what is proposed here (Williams, 2001). Empirical research and theoretical concepts alike could benefit from a combination of literature from governance, programme management, spatial planning with water management, that focuses on the development of integrated plans as well as their implementation in practice.

Chapter 5

Collaborative research to support transition towards integrating flood risk management in urban development

Chapter 5 Collaborative research to support transition towards integrating flood risk management in urban development

This chapter has been published as:

Van Herk, S., Zevenbergen, C., Rijke, J., Ashley, R. (2011). Collaborative research to support transition towards integrating flood risk management in urban development, *Journal of Flood Risk Management*, Volume 4, Issue 4, December 2011, Pages: 306–317

Abstract

Urban development and regeneration present windows of opportunity to help reduce flood vulnerability. Much recent research stresses the need for better integration of flood risk into planning processes. However, there is limited experience in incorporating flood risk management in urban planning. Demonstration projects can help in overcoming this lack of experience and facilitate a transition towards integrated flood risk management and spatial planning; taking longer term flexible and adaptive approaches. This paper demonstrates that the process of research, if collaborative, can contribute to the better integration of flood risk in urban planning. Collaborative research can support demonstration projects and their wider uptake, as well as policy development and subsequently a transition to integrated flood risk management. It also provides greater freedom to use different approaches, consider a broader range of problems and solutions, bring stakeholders together and facilitate capacity building. This paper calls for collaborative research that facilitates social learning and is impact focussed, preferably supporting innovative demonstration projects.

5.1 Introduction

5.1.1 Urban development as a window of opportunity for flood risk management

Worldwide urban floods are increasing in frequency and impact (Munich Re, 2009). Although the projected impacts may be marginal increases on the already large flood losses, in the coming decades climate change will have a major impact on the way in which flooding is dealt with longer term. It is increasingly recognised that engineering alone cannot accommodate the future frequencies and impacts of flooding and a shift in emphasis is required from hard structural solutions to a mixed integrated approach that consists of both structural and non-structural responses. Among the non-structural responses land use planning is considered as one of the most crucial in managing exposure and vulnerability to floods.

Many cities around the world are facing the challenges of sustainable development and are exploring ways to enhance flood resilience. According to White (2008) reflexive, knowledgeable and adaptive cities are needed. It important that all stakeholders recognise that the future is inherently uncertain and that science will not necessarily reduce uncertainty (Neufville, 2004; Klinke and Renn, 2006). Strategies which address these challenges have in common that they recognize that there is no best solution and embrace a future which fits into a distribution of events that will not come as a surprise (Rose, 2004; Pahl-Wostl, 2006). Hence, climate change provides an incentive to think long term and incrementally and consequently to consider reforming traditional urban planning systems and flood management approaches. These systems and approaches should be flexible and adaptable ((e.g. White 2008; Bosher & Coaffee, 2008). Normal urban dynamics of cities provide windows of opportunity over time to adapt and to reduce vulnerability, especially as important drivers include often unplanned, urban developments (Zevenbergen et al., 2008). This is illustrated in figure 5.1. Once an area is developed or regenerated, physical spatial measures can be taken such as: restoring the natural flood plain for storage, unsealing of hard surfaces or creation of green spaces for natural drainage and flood storage, transport networks lowered for flood storage or elevated as evacuation routes, flood proof buildings or green roofs, etc (White, 2008).

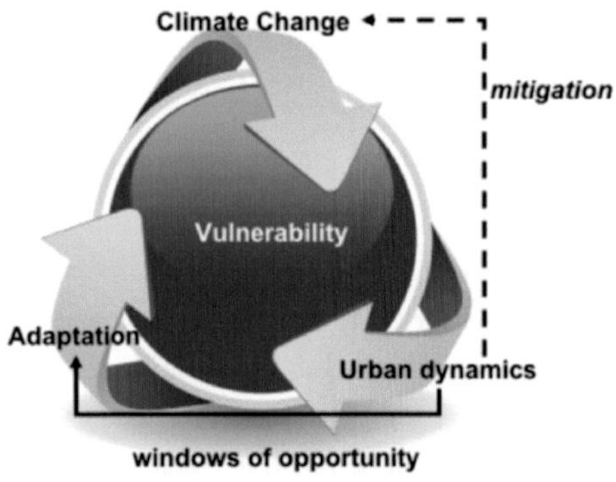

Figure 5.1. Urban dynamics provide windows of opportunity to adapt and reduce vulnerability

5.1.2 Transition to integrated flood risk management

Traditionally an integrated approach to flood risk management and land use planning has not been widespread (e.g. Carter et al., 2005). There are many barriers to integration and implementing coherent approaches. The first group of barriers relate to characteristics of urban development. Urban planning involves integrating many sectoral interests of which flood risk is merely one, and not normally considered to be the most important. The integration of external priorities or interests within planning processes and procedures, the principle of 'external integration' (Spit & Zoete 2006) is common and is part of the general ambition to optimize planning processes at least in the Netherlands. Effective flood risk management relates to more long term and costly public interests as opposed to purely short-term economic interests. A shift to more market led planning in e.g. the Netherlands, which tends to be more short-term profit led, fragmented and locally unaccountable has been found to be inhibiting for flood risk management. Also different Institutional players may have different priorities (Meuleman, 2008).

The probability of incorporating effective flood risk management, that generally has a longer planning horizon than urban planning (respectively 100 years compared with 5-25 years), is even further constrained worldwide where a mere 5% of new development 'under way' in the world's expanding cities is planned at all (Gentleman, 2007), and where no stakeholder has final or total control on urban or spatial developments (Sellers, 2002). The many actors involved further complicate clear and apparently direct attempts to address flood risk, even in developed countries (e.g. Veerbeek et al, 2010). Other barriers for implementation of a more integrated approach include the increased future uncertainty of flood risk (Milly et al., 2008) and a lack of understanding or shared perception of the effectiveness of non-standard response measures (Adger et al., 2005), making flood risk more difficult to plan and manage than other areas of urban planning. The uncertainties also contribute to a technical lock-in into 'standard' solutions such as large structural defence measures (Walker, 2000). The limited experience with incorporating flood risk as a planning variable to reduce exposure and vulnerability, explains why qualitative research indicates a lack of capacity amongst many professionals and politicians and entrapment in traditional cultures and approaches. For example, the Netherlands has a tradition of both flood defence and spatial planning, but these are dealt with separately. Woltjer (2007) points out that there is a move towards strategic spatial planning with water in the Netherlands, but also highlights the impeding institutional conditions, found in many developed countries in relation to water (e.g. Barnett & O'Neill, 2010) potentially risking maladaptation. Even in the Netherlands existing institutions, regimes and policies are designed for the 'fighting the water' paradigm rather than for 'living with water' (e.g. Deltacomissie, 2008), and flood defence is mostly a national public responsibility whereas urban development is a local public-private affair. Similar confusion exists in England and Wales in regard to distancing planning from effective flood risk management, further compounded by private water companies being responsible for urban drainage (Veerbeek et al, 2010).

A transition is necessary to overcome the barriers and effectively incorporate flood risk as a planning variable to reduce exposure and vulnerability. The barriers point towards 'wicked-problems'; being problems that have multiple and conflicting criteria for defining solutions, solutions that create problems for others, and no rules for determining when problems can be said to be solved (Rittel and Webber, 1973) or persistent problems that are ill-structured, involve many stakeholders, are surrounded by structural uncertainties, and are hard to manage (Rotmans, 2005). Hence, Van der Brugge (2005) and Rijke et al (2008) stress the need for a transition towards more integrated water management.

Planning literature (Hajer et al., 2006; White, 2008; Boelens, 2006; Mommaas and Janssen, 2008; Salet et al., 2003) reflects a similar view that the lack of capacity to deal with wicked problems requires active learning. A move is proposed towards more collaborative development planning, spatial governance and organised connectivity. Hajer et al. (2006) express the need for a transition to horizontally, interactive planning, as opposed to vertically institutional. They state that planning regimes should be flexible and dynamic enough to address contemporary, complex challenges, combining spatial quality with democratic legitimacy. However, operationalising the shift from 'government to governance' is difficult as procedures and instruments adhere to government structures in a representative democracy. Seizing windows of opportunity in urban development at a local level to decrease flood risk can address these perceived needs for transition.

It follows from the above that current urban flood management regimes require a transition towards more adaptive and integrated approaches. "The challenge is to provide the means to effect the required change in culture (a transition process) especially amongst the main players and to find new, adaptable and resilient, innovative ways of addressing future security" (Ashley et al., 2012, p2). There is a broad literature on the theory of 'transition' and 'active learning' in the water management domain. Many have argued that social learning mechanisms can promote and support socio-technical transitions (e.g., Pahl-Wostl, 2002; Breit et al., 2003; Keen et al., 2005; van de Kerkhof and Wieczorek, 2005; Gunderson et al., 2006; Ison et al., 2007; Pahl-Wostl et al., 2007). Folke (2006) adds that social learning also helps to build experience to cope with uncertainty and change, which is especially relevant for integrated flood risk management and urban planning. Social learning here refers to both the capacity building of individuals and organisations, and also the creation of relational qualities and social capital (Pahl-Wostl et al., 2007; Bouwen and Tallieu, 2004).

This paper shows that collaborative research projects can be a vehicle for social learning and thus help to deal with uncertainty. Collaborative research naturally focuses on knowledge generation between stakeholders; capacity building and creation of relational qualities. If knowledge uptake is also managed effectively, collaborative research can contribute to a transition. This paper provides an example of pathways for collaborative research that have supported a transition based on the Dutch project Urban Flood Management Dordrecht. Recommendations are also provided as to how best to use such pathways effectively."

5.2 Evaluating collaborative research to support transition

In Europe, there is a large body of flood related research. Not including European research programmes such as the Framework Programmes, national research in 13 European countries represents an investment of 55M Euro/year that is between 1-4 % of the total investment in flood risk management (ERA-Net CRUE, 2007). The portfolio of these research programmes covers: transdisciplinary and disciplinary research, research on natural, technical and socio-economic sciences, and 'science' and 'policy' oriented research, yet with a different focus for each country (ERA-Net CRUE, 2007). Many of these research programmes, especially the transdisciplinary, socio-economic and policy oriented research aim to support a transition in flood risk management. These include the Dutch Living with Water (LwW) programme that supported the case study, the collaborative research project Urban Flood Management (UFM) in Dordrecht. The LwW programme is highlighted in figure 5.2, the highlighted ellipse in the top-right as policy oriented, trans-disciplinary and socio-economic research.

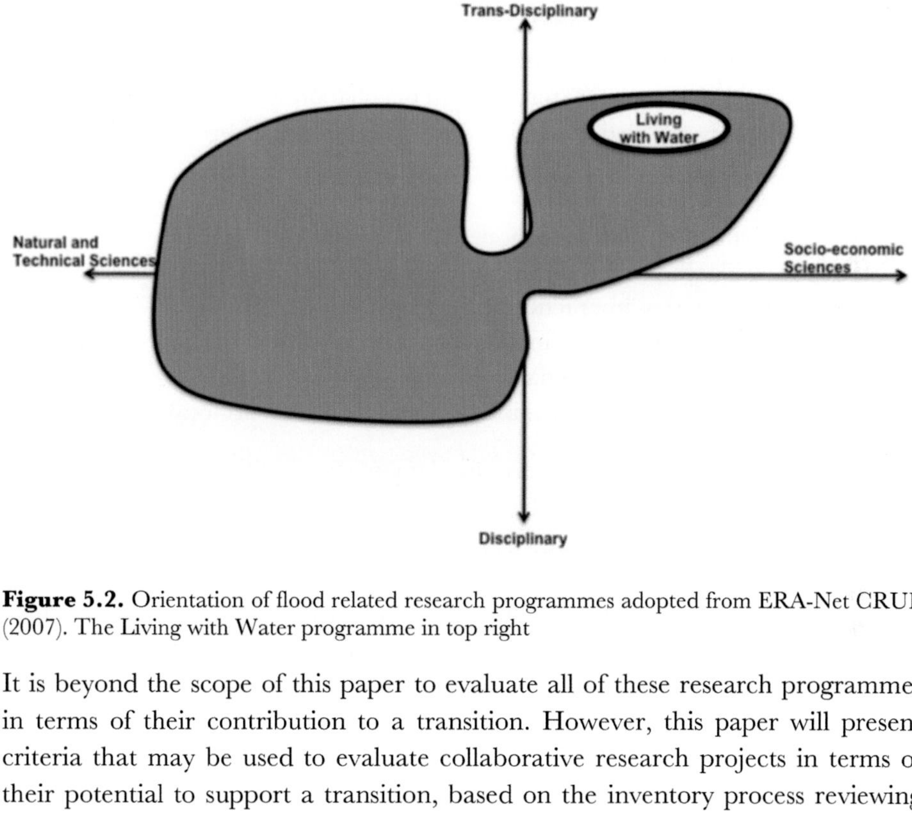

Figure 5.2. Orientation of flood related research programmes adopted from ERA-Net CRUE (2007). The Living with Water programme in top right

It is beyond the scope of this paper to evaluate all of these research programmes in terms of their contribution to a transition. However, this paper will present criteria that may be used to evaluate collaborative research projects in terms of their potential to support a transition, based on the inventory process reviewing the pathways for collaborative research used to support the transition for the UFM Dordrecht case study. This in turn, can help others to answer the questions: are we 'doing the right research' and are we 'doing the research in the right way' where a transition from one FRM regime to another is expected to be part of the outcome.

Existing definitions and reasons for collaborative research are presented in Katz & Martin's work: what is research collaboration. They focus on collaboration in science and not social science and do not consider social learning and transitions. This study may be too narrowly defined for the main purpose of the research presented here. However, it is nonetheless possible to build upon this body of knowledge in combination with recent definitions of social learning frameworks. Katz & Martin (1997, p11) define research collaboration: "*as the working together of researchers to achieve the common goal of producing new scientific knowledge*" and they distinguish research between: individuals, groups, departments, institutions, sectors and countries. Cornwall & Jewkes (1995) view collaborative research more as an attitude or approach rather than well-defined techniques to be followed, and highlight that participatory research - a form of collaborative research - focuses on sequential reflection and action carried out with and by local people rather than on them. Katz & Martin (1997) present 6 reasons for collaborative research:

- Funding agencies need to save money
- Growing availability and falling cost of transport and Communications
- Desire for intellection interaction with other scientists
- Need for a division of labour in specialisations
- Requirements of interdisciplinary research
- Government encouragement of international and cross-sectoral collaboration.

These ideas focus mainly on more fundamental scientific research projects, researchers and their efficiency. These may be considered as the 'providers of knowledge'; in particular explicit, factual and impersonal knowledge. However, these theories do not seem to consider: knowledge as socially construed; normatively loaded reality definitions and images (Schön & Rein, 1994, Fischer, 1990); or knowledge as experience-based competencies and skills (Cook & Brown, 1999). By collaborating or otherwise interacting with non-scientific stakeholders the latter types of knowledge are further developed and can be applied better in policy and daily practice. Here, collaborative research projects and researchers that focus on these types of knowledge can be considered either: 'messengers of knowledge' collaborating directly with practitioners and policy makers or 'ambassadors of knowledge', where leading scientists become advisors to policy makers. These types of 'researchers' are interconnected as illustrated diagrammatically in figure 5.3. Similarly, Pielke (2007) distinguishes between roles such as 'the pure scientist'; 'the science arbiter'; 'the issue advocate' or 'the honest broker of policy options'.

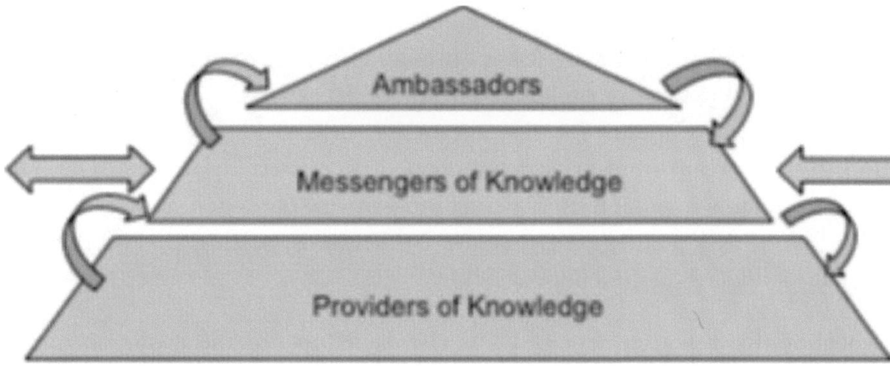

Figure 5.3. Diagrammatic illustration of 3 types of scientists: ambassadors, messengers and providers of knowledge

For the purposes of this paper it is necessary to consider a broader group of collaborators than 'researchers' and a broader range of results than just 'scientific knowledge'. In order to contribute to a transition, end-users, practitioners, policy makers and many more stakeholders need to be involved to support e.g. the development of new, innovative policies. If knowledge is broadly defined as above, knowledge generation and also knowledge uptake have a pivotal role in collaborative research projects supporting a transition. The case study may therefore be used as a collaborative research project which has a main impact via a transition obtained at the level of 'messengers of knowledge'.

5.3 Case study: the impact on transition of urban flood management in Dordrecht

The Urban Flood Management (UFM) Dordrecht research project (2005-2008) was set up by a consortium of public, private and research partners from local to national level to learn about flood resilient and climate proof urban development in areas outside the main flood defences. UFM used the redevelopment project De Stadswerven as a case study; a former shipyard area, located on the edge of the historical city centre of Dordrecht, the Netherlands. The UFM research project radically altered the approach being taken from one of 'flood prevention' to that of 'flood risk management' in common with initiatives elsewhere (e.g. Evans et al., 2004). The original concept for the De Stadswerven project followed the focus on flood prevention set by national government; specifying the raising of the site ground level from an average of 2-3m to 4m above mean sea level. Four years after inception of the original plan, an *integrated* flood risk management approach was adopted, i.e. flood risk management within the planning system; modifying the masterplan to include a range of alternative flood risk reduction measures such as: floating buildings; flood resistant measures at the lower building levels; and elevated roads and walkways providing escape routes for evacuation. The UFM research project provided the flood risk analysis and management approach, and the impetus for the alternative urban designs and capacities developed by a consortium that championed a revised approach. The De Stadswerven plan is now seen as a leading innovative example of how to develop integrated flood risk management in the Netherlands, seizing the window of opportunity spatial development provides.

The collaborative research project UFM clearly influenced the evolution of its' own case study, a redevelopment project in Dordrecht. Moreover, stakeholders indicated that UFM actually had a much broader impact on the transition from a 'flood prevention' to a 'flood risk management' culture. Sixteen semi-structured interviews and two workshops were used to ask UFM participants, project leaders of De Stadswerven and policy makers about the impact of the project in relation to this transition. The workshops aimed at data collection, but the researchers had different roles as observer, analyst and facilitator, meaning they inadvertently became involved in participatory action research, a research method in its own right (Robson, 2002).

From the analysis and data triangulation of interview and workshop outcomes, six clusters of apparent 'impact factors' emerged that were believed by the stakeholders to have supported the transition.

1. Science. There is now a new and improved method to quantify flood risk. A scientific advance from UFM was the development of new methods for flood risk modelling in the Netherlands: fine grain models (Veerbeek and Zevenbergen, 2009). These models enable detailed flood damage assessment, specified for buildings, neighbourhoods, and types of damage. In terms of supporting a transition, this method can support flood resilient urban planning and design and can support political debate on flood safety standards. This quantitative method is considered pivotal in the public-private debate on the potential use of flood risk insurance for such areas.

2. Practice and demonstration projects. The new masterplan for the De Stadswerven redevelopment (practice) ultimately incorporated several flood risk reduction measures designed in the UFM project. It included 'flood free routes' and water-rich areas where the tidal influence can be lived with by inhabitants and areas that can be intermittently flooded and thus help to offset increased risks from climate change. A pilot project for 100 flood-proof dwellings was initiated to experiment with the concepts before the new masterplan was adopted. This innovative project is a demonstration of new practice that can inspire other development projects in flood-prone areas. UFM design concepts are currently being used by the city of Rotterdam for the redevelopment of a harbour area, thus providing a transition beyond the spatial confines of the development itself. Transition theory (see also Rotmans et al. 2001; Kemp et al., 1998) recognizes the importance of such demonstration projects in making large scale transitions happen.

3. Policy. UFM showed that urban development in areas outside protective dykes does not necessarily compromise flood safety. From this insight the Province of South Holland was able to develop a new and more permissive policy for developments in such areas. The UFM findings were also adopted by the national Delta Commission for the new flood safety policy in The Netherlands. As these research results are being adopted into policy they are 'scaled up' and influence practice more broadly.

4. Awareness. The UFM research project and its results were extensively publicised to stakeholders and the general public. Media coverage in amongst others: Dutch, French, British newspapers and television supported the dissemination. Also the visits of the Dutch crown prince and the national committee for Delta Technology brought the newly developed approach to the attention of more stakeholders. This, in turn, increased awareness of the possibilities to develop urban areas outside protective dykes using an integrated flood risk management approach. Farrelly and Brown (2008) have used the term *receptivity* as a measure for the chances for uptake of new approaches or technology, based on Jeffrey & Seaton (2003/4). If awareness is created amongst users the receptivity increases which is fundamental for any transition.

5. Network. UFM itself comprised of a network of stakeholders working together in this research project. The UFM consortium decided to maintain the network and to set up a continuing research project. In the new project, they have formed a so-called Learning and Action Alliance to support new, demonstration projects and other policy challenges, such as green infrastructure and multi level safety (Ashley et al, 2010). Networks are a form of relational qualities obtained by social learning and crucial for transtions (Pahl-Wostl et al., 2007; Bouwen and Tallieu, 2004). Hillier (2000) stressed the importance of networks in planning processes.

6. Capacity building. Farrelly et al (2009) explain further how local-scale experiments can be valuable learning platforms for urban water practitioners. In the UFM Project all those surveyed (see above) recognized that they had learnt and developed new competences because of the project and their involvement in the practical case study De Stadswerven. The competences developed differed between participants and included: 1. single-loop or instrumental or technical learning; 2. double-loop learning, adjusting their general frameworks of beliefs, norms and objectives; and 3. deutero-learning or learning how to learn (Tuinstra, 2008).

It may be concluded from the descriptions of the six impact factors above that the types of impact are related or may be mutually dependent. Capacity Building is at the focus of these factors, as are networks. Awareness is needed for all of the factors. Science, Practice and Policy influence each other.

5.4 How collaborative research can support transition

From the case study above it is possible to conclude that collaborative research can be defined as potentially impacting on a transition via six factors whereas Geels and Schot (2007) distinguish three levels of socio-technical transition pathways: Macro, Meso and Micro (figure 5.4). The lessons from the case study here indicate that collaborative research of the type undertaken can act on two of these transition pathways: the socio-technical regime (Meso level) and Niche Innovations (Micro level) (Figure 5.4). Landscape developments (Marco level) can put pressure on the regime creating windows of opportunity for transition. Rijke et al. (2008) highlighted the following pressures for the water sector in the Netherlands: environmental awareness, a disruptive change since the 1960s; climate change, a disruptive change since the 1990s; and flood events, shock changes in 1993 and 1995. Recent flood events or even the financial crisis (2008) can result in other shock changes. Research is not able to influence such 'landscape' developments, but can recognize and support the exploitation of windows of opportunities on the meso and micro level. On the meso level research can actively catalyse transition through supporting the creation of new, interdisciplinary partnerships and networks. The networks can be used to influence policy, to support capacity building and dissemination for broader practice and to create awareness. On the micro level research can support innovative practical demonstration projects. Figure 5.4 illustrates the role of collaborative research in transition pathways towards integrating urban development and flood risk management.

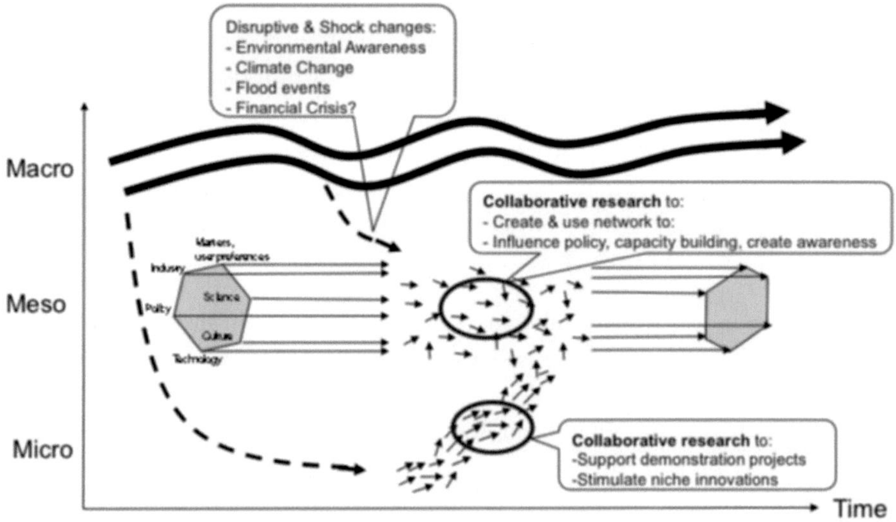

Figure 5.4. The potential role of collaborative research in the transition pathway (adapted from Geels and Schot (2007) and Rijke et al (2008)

Using the lessons from the case study and the earlier theory presented, it is possible to stress at least one criterion that is needed for the design and evaluation of collaborative research projects and portfolios related to FRM and urban planning. 'Doing the right research' means having collaborative research that supports practical, innovative projects that combine urban development with flood risk management in the flood related research portfolio. Thus, research is required using and supporting an actual case study ex-ante, prior to development, or ex-durante, during the urban development project. Such research would directly support the micro level of the transition pathway (Figure 5.4). Schön (1983) stated that demonstration provides opportunities to effectively 'learn from experience' by 'learning by doing'. All of the UFM stakeholders surveyed in this study were found to have had their motivation increased and a significant learning experience by working on this practical, real-life project.

Another criterion 'doing the research right' relates to the collaborative network and thus social learning component. Collaborative research can be a form of social learning framework. Social learning mechanisms or the frameworks of multi-party collaboration necessary to deliver social learning and stimulate interactive decision making have been given many names, such as: Communities of Practice (Wenger, 2000), Learning Alliance (Verhagen, 2008; Batchelor & Butterworth, 2008), Learning & Action Alliance (Ashley et al., 2012), socially-embedded institutions (Cleaver, 2002), learning platforms or arenas (Farrelly et al., 2009). The theory of transition pathways shows how networks, developed through collaborative research as social learning frameworks, can support regime changes e.g. via policy processes. This is also called a transdisciplinary transition process (e.g. Scholz, 2000; Scholz & Tietje, 2002; Klein, 2001; and Gibbons, 2001). The UFM Project clearly influenced new policy processes and experimented with new practical approaches to flood risk Management for areas outside the protected dykes directly through the collaboration of project participants and partner organisations.

Any collaboration should explicitly aim for an impact on the transition process via inclusion of the 6 impact factors: science, practice, policy, awareness, networks, capacity building. A collaborative research project should have sufficient flexibility to seize opportunities to do this, as opposed to strictly complying with a project plan or with constricting regulations and standards. In fact, it should lead the way in challenging and setting new standards and regulations.

Despite the various definitions above, it is likely that different social learning frameworks are necessary at different stages of a transition. This can possibly explain the various names and definitions for these frameworks. Figure 5.5 shows this hypothesis depicting several social frameworks in the S-curve going through the four transition phases: pre-development, take-off, acceleration, stabilisation (see also Rotmans et al, 2001). Collaborative research would provide an appropriate framework during the early stages of transition: pre-development and take-off phases (Figure 5.5). From this, innovative approaches are being developed and experimented with preparing the way for the acceleration phase by creating networks and addressing policy and regulatory problems and finding solutions to these. In the final stage, stakeholders still have to work together and learn together, for example to deal with uncertainty, but this is a more continuous need and activity. Frameworks such as Communities of Practice and socially-embedded institutions seem more appropriate at this stage. Literature on social learning is not clear in the definitions of the frameworks proposed regarding their role in a transition, if any. It would be useful to position these previously defined social learning frameworks within the larger transition picture (Figure 5.5). As a suggestion, the figure illustrates the likely timing of collaborative research and other social learning frameworks, which are not mutually exclusive, according to the state of transition they might best contribute to. Further work is needed to develop and analyse social learning frameworks in terms of these different phases.

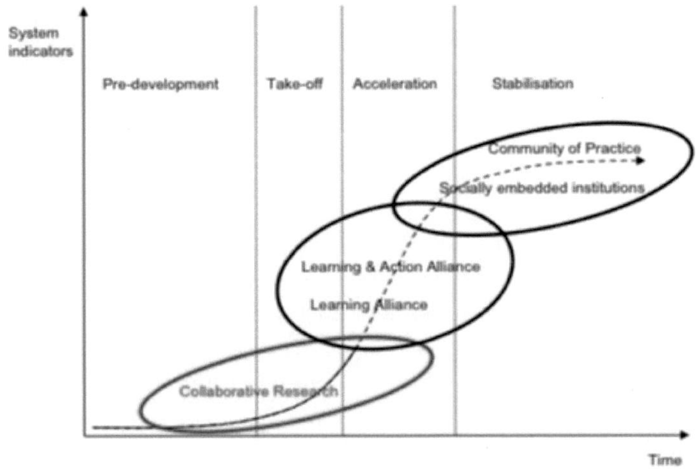

Figure 5.5. Timing of collaborative research and other social learning frameworks according to the state of transition they might best contribute to. (adapted from Rijke et al.,2008 on state of transitions)

5.5 Collaborative research supporting urban development and policy making

Hence, from theory and practice it is found that collaborative research can support a transition via the 6 factors and 2 transition pathways. The question arises how to effectively manage the 6 impact factors. This section focuses on two of the six impact factors, policy and practice.For collaborative research to support urban development projects and policy making, this firstly requires an understanding of both processes and related complex decision making. This section presents some models to understand these processes, whilst the next section presents factors that can support the uptake of research results in these processes.

Decision makers have to balance multiple objectives, e.g. flood safety which should be combined or can compete with housing, ecology, economy, and many other objectives. Logically all these objectives mobilise multiple stakeholders, each with their own interests. There is not one single decision maker. Therefore, Healey (1998) recommends a relational view of governance processes instead of public action as the delivery of products by specific agencies. There are many stakeholders with decision-making power in, and from different policy fields and at different levels, public and private. Other stakeholders can have blocking power, others are simply affected by the problems or measures. Additionally the stakeholders' perspectives of problems and ideal solutions change over time, which introduces additional dynamics to the decision-making process.

Teisman (2000) presents three conceptual models to analyse decision making processes that are relevant for urban development as well as for policy making in a broader sense: the phase model, the stream model and the rounds model.

- The phase model (Mintzberg et al., 1976) focuses on successive and distinctive stages in a process, i.e. defining a problem, formulation, adoption, implementation and evaluation of a policy or solution.
- The stream model (Kingdon, 1984) emphasizes concurrent streams of participants, problems and solutions, defining decision making as the connection between these streams.
- The rounds model combines elements of the other two models, in assuming that several actors introduce combinations of problems and solutions and create progress through interaction.

The three different models show that it is not always obvious that policies are set at a certain moment by a certain actor. Policies result from a series of decisions taken by various actors. Also policy evaluation is not that straightforward. Compliance with the objectives of all parties needs to be evaluated.

These models provide conceptual insights into how decision-making can be supported by collaborative research. Following the stream model, research can: (i) address problems; (ii) propose solutions; and (iii) bring participants together. Thus pushing and pulling the streams to take advantage of or even provoke a (policy) window of opportunity. Also timing is everything. In the case study De Stadswerven it was found that the collaborative research project UFM redefined the problem of flood risk and proposed flood-proof planning and building solutions that were considered appealing to decision makers. The De Stadswerven project was temporarily stopped due to political tensions, allowing UFM to develop and subsequently enhance political receptivity for the new integrated flood risk management approach. Similarly, related regional and national policy processes were amenable to external and new ideas at the time and were able to use the insights obtained in UFM.

When using the insights gained from the phase and round models, it is apparent that on-going research should understand and take cognisance of the round or phase in the decision process. As the phase and round models are useful for (ex-post) analysis, it is difficult to be able to define the round or phase at any given moment during the research process. It is necessary to use 'sensors' or key signals for step-changes in the process, such as a change of the decision makers after elections. Personal contacts are good sensors. It is necessary to understand their interests and drivers in order to bring participants together or to address problems or propose solutions. Any stakeholder who wishes to influence the definition of reality (e.g. ideas on problems, solutions, actions, objectives) of the other stakeholders in a policy process, will only succeed to the extent that they are able properly to understand the 'definition of reality' of these other stakeholders (Termeer and Koppenjan, 1997).

Knowledge can be used in decision making in different ways. Van Buuren (2006) presented an idealized description of competent decision-making, using knowledge, in three interrelated threads: (1) thread to *establish facts*; (2) thread to *create images*; (3) thread to *set ambitions*. As knowledge, in a broad sense, is the main output of collaborative research, it is to be expected that collaborative research can support the fact (1) thread above by generating factual knowledge, the image (2) and ambitions (3) threads by bringing stakeholders together in a way that they can freely discuss interests and views.

In short, practice and policy of flood risk management and urban planning involve interactive processes with multiple stakeholders and interests. Collaborative research can help organise multi-actor collaboration by providing a more informal, non-threatening environment for actors to discuss problems, solutions, facts, perceptions or images and ambitions. The depth of participation can vary from: Informing, consulting, advising, coproducing to co-deciding (Edelenbos and Klijn, 2006). Likewise, collaborative research can support collaborative planning by bringing policy fields together to develop integrated solutions for multiple ambitions. A forthcoming paper uses the models presented in this section to analyse two case studies and to draw lessons on how to effectively organise social learning (Van Herk, 2011a).

5.6 Research impact: receptivity and advocacy

The decision-making processes for urban development and policy development have their own dynamics of problems being addressed, solutions being proposed and stakeholders participating. These processes can be supported by collaborative research, often being an independent process. This also applies to the supportive role of research in a transition towards integrating flood risk management in urban development as mainstream accepted practice. To organise collaborative research to effectively support this requires knowledge of the factors that the uptake of knowledge or research results in urban development projects, policy and wider practice depend upon.

Farrelly and Brown (2008) utilised the term *receptivity,* considering that a new technology or initiative must be designed from the end-user or recipient's point of view. Thus, research should be designed from the point of view of the involved persons in targeted urban development projects and policy-making trajectories. Farrelly and Brown present 4'A' attributes of receptivity:

- Awareness: individual or organisation is aware of a problem and need for a solution.
- Association: individual or organisation relates to the potential benefits, enough to expend effort to apply solution(s).
- Acquisition: individual or organisation has requisite skills, capacities and support to implement solution(s).
- Application: incentives are available to encourage the individual or organisation to implement solution(s).

Thus, individuals and organisations participating in collaborative research projects can influence the receptiveness. They can support the creation of awareness by addressing problems and solutions. They can support association by presenting potential benefits with a scientifically sound justification. They can also support acquisition through capacity building, involving individuals and organisations in the learning environment and can advocate development of e.g. policy and regulation to create incentives for application. Thus, the receptivity attributes point towards the impact factors found in the case study: awareness, capacity building, networks and policy.

In parallel with 'receivers' and their receptivity, this paper introduces collaborative researchers as 'senders' or 'messengers of knowledge' and introduces: *'advocacy', a fifth 'A'.* This is senders' efforts to 'advocate' their knowledge to be taken up in applications. Thinking in terms of 'advocacy' provides collaborative research with a perspective for action to influence the receptivity of the urban development projects or policy processes. Thus, in this context, this paper defines *advocacy* as: the support of collaborative research to apply its knowledge in demonstration projects and policy processes by generating and exploiting receptiveness in these projects and processes.

Following the concepts of receptivity and advocacy above, there is a further conceptual framework that may be used to support the uptake of knowledge. The framework is adapted from and inspired by communication theory (e.g. Schramm, 1982) and technology transfer theory (Bozeman, 2000) as knowledge has to be 'communicated' or 'transferred' to urban development or policy making. In this paper participants in collaborative research are termed: 'messengers' or 'ambassadors of knowledge'. This paper introduces: senders, recipients, message(s), medium and support environment. Senders should understand receptiveness to more effectively choose the message, medium and the individual receiver. They should also speak the 'language' of receivers and be well connected to receivers in the other processes. For this, networks should be created and exploited and ideally some senders should also be receivers. Messages contain knowledge which content and form should be aligned with the needs and language of the receiver. E.g. in the case study UFM the message included spatial design alternatives that reduce flood risk and are considered attractive living environments. Effective mediums are carefully selected or created, carefully timed and involve as much direct personal contact as possible between sender and receiver. Collaborative research projects could present many important mediums. For example, in the case study UFM, a pilot project for 100 flood proof dwellings was initiated. The solutions proposed for the pilot (message), were well-received by the development consortium (receivers) that worked together with the researchers (senders) in this initiative (medium). The national, public-private committee on delta-technology further supported this initiative (support environment).

From the case study it is clear that people are an important success factor for the impact of collaborative research for a transition. Many authors stress the importance of people as knowledge brokers (Pielke, 2007) or champions (Taylor, 2008). In this study the stakeholders surveyed highlighted several characteristics of people that were important in creating impact or generating and advocating knowledge. We have clustered their *characteristics* into *6 'C's*, their: competences, commitment, connectedness, collaborative attitude, communicative, and their continuity in the collaborative research, urban development and policy processes. Communicative people with better contacts were more effective in advocating knowledge. Whereas the rotation of participants, leaving for other projects or jobs, reduced the impact of the collaborative research. Stakeholder competences, commitment and collaborative attitude were also important.

5.7 Conclusions

Collaborative research can play an important role in supporting a transition towards integrating flood risk management in urban development processes. Doing the right research and doing the research right is key. This paper advocates research to aim to support practical, innovative urban development projects and create and use networks to influence policy, broader practice, capacity building and awareness. For this, research should have a truly collaborative character promoting active learning and co-production. If partnerships are durably organised in social learning frameworks they can stimulate continuous learning, necessary given the uncertainty regarding climate change and the continuous challenges of urban dynamics. Collaborative research can be a first step towards establishing and sustaining such frameworks. Moreover collaborative research can be a tool for multi-actor collaboration by providing a more informal environment to discuss problems, solutions, facts, perceptions and ambitions amongst a wide group of stakeholders in a neutral and legitimate environment. As such, this can support collaborative planning by jointly developing integrated spatial solutions for multiple ambitions.

It is crucial to understand the underlying complex decision making processes for collaborative research to be effective in supporting practice and policy making. The receptivity to a collaborative research project's knowledge, approaches and partnerships varies in time and across stakeholders. Problems and solutions should be addressed timeously and to the right stakeholder to tap into that receptivity. This paper presents several pathways to increase the impact of research collaboration. More case study analysis of collaborative research supporting practice and policy such as presented here is needed to draw out and verify lessons on effectively organising collaborative research to support a transition. This paper indicates several success and inhibiting factors based on one case study.

Chapter 6

Understanding the transition to integrated flood risk management in the Netherlands

Chapter 6 Understanding the transition to integrated flood risk management in the Netherlands

This chapter has been published as:

Van Herk, S., Rijke, J., Zevenbergen, C., Ashley, R. (2013). Understandiong the transition to integrated flood risk management in the Netherlands. *Journal of Environmental Innovations and Societal Transitions (EIST).* doi: 10.1016/j.eist.2013.11.001.

Abstract

The Multi-Pattern Approach (MPA) is a new method that has been applied to understand the transition to integrated flood risk management (IFRM) in the Netherlands. This paper presents a detailed analysis of the outcomes of the 2.3 billion Euro flood safety programme Room for the River (RftR). 2 years of research, 55 interviews, a survey of 155 respondents and elaborate document analysis, provided in-depth evidence of how the transition occurred in practice. Experiments were scaled-up, IFRM was consolidated in national policies, and the implementation of RftR further adapted the functioning of the societal system. Lessons are drawn that enrich the MPA framework and that can help its further development and application. The MPA provides scientists with a method to analyse transition dynamics as a chain of patterns that occur under certain conditions. Policy makers can shape and monitor the outcomes that are to be generated to support a transition.

6.1 Introduction

The frequency and consequences of extreme flood events have rapidly increased worldwide in recent decades (e.g. Bouwer et al., 2007; Kron, 2009) and climate change is likely to exacerbate this trend in the near future (e.g. IPCC, 2007). The key factors for this increase in flood risk are global population growth and the increase in socio-economic activities in flood prone areas, together with their growing interdependency on flood protection and drainage infrastructure of which a significant part is of unknown or poor condition (Ashley and Cashman, 2006; National Committee on Levee Safety, 2009). It is increasingly recognised that engineering responses alone cannot accommodate the future frequencies and impacts of flooding and a shift in emphasis is required from hard structural solutions to a mixed integrated approach that consists of both structural and non-structural responses (Zevenbergen et al., 2008).

Many scholars call for a transition to policies that actively manage flood risk to reduce flood impacts and accommodate floods: 'living with water', rather then a mere focus on flood protection: 'fighting against water' (e.g. ibid; Zevenbergen et al., 2013b; Newman et al., 2011; White, 2010; Dawson et al. 2011). In the recent past, major flood disasters have acted as drivers for changing flood risk management policies (Mauch, 2009). The rethinking of and change in the traditional approach is included in integrated flood risk management (IFRM) policies such as: the EU Flood Directive (EC, 2006); the source-pathway-receptor framework as used by the Environment Agency in England and Wales (EA, 2000); and the multi-layer safety approach in the Netherlands (V&W, 2008). Many experiments, such as research projects, policy pilots, and demonstration projects have been conducted based on IFRM (e.g. Farrelly and Brown, 2011; Van Herk et al., 2011a; Hegger et al., 2012; van den Brink et al., 2013). An integrated approach to flood risk management set within land use planning processes is now seen as an effective way of minimising flood risk, although this has not always been recognised in practice and implementation is often still lacking (e.g. Watson and Adams, 2010; DCLG, 2012). Implementation of IFRM faces multiple barriers such as: technical lock-in to structural solutions (Walker, 2000) such as defence measures; lack of understanding of the effectiveness of non-standard response measures (Adger et al., 2005). Also spatial planning has many drivers other than flood risk that do not always give IFRM priority consideration (Van Herk et al., 2011a). A transition or regime change is required to overcome these and other barriers (e.g. Van der Brugge and Rotmans, 2007). The implementation of new integrated policies will have to address these and possibly other as-yet unforeseen barriers. We pose the hypothesis that investment programmes that embrace a new

IFRM approach would contribute to a transition to IFRM. By their implementation, such programmes would not only deliver on their objectives (output), but would also generate outcomes that have an impact beyond the scope of the programme and are sustained after the delivery of the programme. These programmes need to be monitored and evaluated to document and draw lessons on the impact of these processes in supporting a wider transition. Those managing, or rather, contributing to a transition, need to understand these processes for iterative adjustment of the governance practices that are needed for such transitions (Loorbach, 2007).

This paper uses as a case study the transition to IFRM in the Netherlands as described by e.g. Van der Brugge et al. (2005) and focuses on the large-scale implementation programme Room for the River (RftR). The paper presents results from 2 years of case study research that aimed to analyse RftR's contribution to a transition to IFRM in the Netherlands through the generation of outcomes. A detailed analysis of the outcomes of RftR is used here to provide in-depth evidence of how the transition occurred in practice during the implementation of the RftR programme. RftR was selected here as a case study, because it is an exemplary project for IFRM in terms of integrated outputs and collaborative processes (Rijke et al., 2012b) and because RftR was positioned as an iconic project in the transition in Dutch water management (Warner et al, 2012). RftR was launched when the dominant flood management paradigm was shifting from 'flood defence' with a sectoral and technological focus, to an integral and spatial focus (Zevenbergen et al., 2013a; 2013b; Van der Brugge et al., 2005). As a response to the floods of 1993 and 1995 in the Netherlands, the 2.3 billion Euro flood safety programme RftR was approved by Dutch parliament in 2006 to increase flood safety by giving the rivers in the Netherlands more room instead of merely reinforcing the defence systems (PKB, 2006). The programme is to be delivered by 2015 to increase the river discharge capacity to 16.000m3/s in Lobith (where the Rhine crosses the German-Dutch border) by implementing river widening measures. The Programme comprises of 39 measures or projects for giving more room for the rivers Rhine, IJssel, Waal and Lek. The concept of river widening comprises measures, such as flood by-passes, excavation of flood plains, and dike relocation. 'Room for the River' explicitly aims to increase flood safety combined with increased spatial quality of landscape, nature and culture (Schut et al., 2010).

The Multi-Pattern Approach (MPA) (De Haan and Rotmans, 2011) is a new method to describe and understand the dynamics of societal transitions as a sequence of patterns (Section 6.2.2). The MPA is applied here for the first time to the transition in the Dutch water sector and has provided interesting insights on the transition, whilst lessons have been drawn on the use of the approach itself. The research questions are: 1. What outcomes have arisen from RftR that have an impact beyond the scope of the programme and are sustained after the delivery of the programme? 2. How has RftR contributed to a transition to IFRM in the Netherlands and can this be explained by the outcomes that have arisen? 3. What specific insights were gained through the use of the MPA and what lessons can be drawn on the use of the approach itself? Empirical evidence is analysed as to what outcomes have arisen from RftR (Section 6.3). The MPA is used to explain the outcomes RftR has generated and how these contributed to a transition to IFRM (Section 6.4.1). The case study as presented in this paper can help the further development and application of the framework and future comparative studies on transitions in the water domain or with other domains (Section 6.4.2). The case study can also be revisited to enrich transition management literature. If project outcomes can explain a project's contribution to a transition, they provide a potential lever for policy makers and project managers. The Delta Programme is the new large IFRM programme in the Netherlands that works on flood safety and fresh water supply both for now and into the future (Deltacommissaris, 2011). It builds upon lessons of RftR (Kabat el al., 2009) and introduces new approaches such as 'adaptive delta management' that combines long term planning with short term action (Deltacommissaris, 2011). This paper documents part of the legacy of RftR in terms of the transition to IFRM and provides useful lessons for the Delta Programme to prepare for, influence or understand the transition (Section 6.4.3).

6.2 Research approach

6.2.1 Methods to study outcomes of RftR

A mixed-method case study approach was used to analyse the outcomes of RftR. 55 face-to-face semi-structured interviews were conducted that covered the following similar topics of relevance for this paper: examples of outcomes of RftR (if any); how these outcomes have an impact beyond the scope of the programme and are sustained after the delivery of the programme (e.g. does the interviewee use these outcomes outside RftR?); why and how these outcomes have arisen. The interviews were conducted with people who were involved in the initiation (n=10), design and realisation (n=31) stage of RftR, as well as people in the initiation stage of the Delta Programme (n=3) and other strategic positions at the levels of senior policy maker and decision maker (n=11). Interviewees represented a range of different disciplines and organisations, including: the Room for the River programme directorate; Rijkswaterstaat (the executive branch of the Ministry of

Infrastructure & Environment); waterboards; provinces; municipalities; the Ministry of Infrastructure and the Environment; and independent scientists and representatives of planning bureaux. Interview transcripts were analysed using QSR Nvivo 9. Data were coded inductively for outcomes of RftR that emerged from the data. In addition to the interviews, examples of outcomes were obtained through observation at: 3 training sessions which combined had 45 participants; 2 political conferences with approximately 110 participants each; one community-building event of RftR and the Ministry of Infrastructure & Environment with approximately 150 participants. The outcomes mentioned by interviewees or observed at events were verified through document analysis of: guidelines (e.g. RVR, 2008; Provincie Overijssel, 2007), policy & regulation (e.g. RWS, 2010), training documents and meeting minutes, progress reports (e.g. PDR, 2011a;b).

After a year of research and the first round of interviews, a survey was conducted of participants in RftR. There were 151 survey respondents: 48 from the RftR programme directorate; 10 from other parts of Rijkswaterstaat; 10 from the Government Ministries involved; 11 from Provinces; 22 from waterboards; 36 from municipalities; 7 from the private sector; and 7 other respondents, such as scientists and community groups. The outcomes that had already emerged from the data were incorporated into a set of questions to quantify RftR's impacts on various changes: river widening as a possible solution; new approaches for design and analysis; collaboration between authorities; project and programme management and organisation. Respondents were asked to rate RftR's influence on these changes using a five-point Likert-based scale, where 1 was very small influence; and 5 very large influence. Also they were asked to indicate if they use the lessons of RftR that are related to these changes in their own organisation or other projects. The survey responses were analysed statistically using IBM SPSS 19.0. The data were analysed for classifications per type of organisation (e.g. national government or municipality), or type of position (e.g. politician or professional).

6.2.2 Multi-pattern approach (MPA)

The multi-pattern approach (MPA) of De Haan and Rotmans (2011) has been applied to describe the transition in the Dutch water sector and analyse the contribution of RftR to this transition. Compared with other frameworks such as the management and transitions framework (e.g. Pahl-Wostl et al., 2010), MPA may be used for studying the dynamics in a transition due to both internal and contextual factors (Ferguson et al., 2011). For the purpose of this research, a dynamic perspective can help to highlight smaller steps such as the contribution of a single programme in a wider transition. The contextual factors have been of paramount importance to the transition in Dutch water management and to RftR (Van der Brugge et al., 2005), whilst internal factors (learning) have also played a role in RftR's contribution to the transition (for an in-depth analysis of learning in RftR, see: Van Herk et al., 2013c; in press).

The multi-pattern approach has been applied to the Dutch water sector, based on a literature review (e.g. Van der Brugge et al., 2005) and interviews with national representatives involved at the inception of RftR. Their descriptions are summarized to understand the historical background of the RftR programme (Section 6.3.1). The outcomes of RftR that were found have been related to transition patterns and conditions that have occurred prior and during the implementation of RftR. Interviews with representatives of future programmes and national policy makers were used to validate the uptake of outcomes from RftR and to discuss new transition conditions and potential future transition paths (Section 6.4.3).

The MPA is based on the prior development of a variety of transition theories with the aim to develop a coherent and integral approach based on a reframed conceptual language. Dating back to innovation studies (e.g. Levinthal, 1998) and research on technological and socio-technical transitions (e.g. Geels, 2005), MPA considers transitions as complex problems that are to be studied from a systemic perspective. A societal system is a part of society that can be attributed a functioning to meet a societal need. A transition is a fundamental change in the structures, cultures and practices of a societal system, profoundly altering the way it functions (e.g. Rotmans and Loorbach, 2006; Rotmans and Loorbach, 2009; Van Raak, 2010). MPA redefines this based on the regime-niche language from social-technical transition studies (Rip and Kemp, 1998). A social system is subdivided into constellations (sub-systems) and the constellation that dominates the functioning of the system is denoted as the regime; the functioning of the regime is the typical way societal needs are met. Niches are powerless constellations with novel, or deviant functioning that can meet quite specific societal needs, often in unorthodox ways. The redefinition of societal transitions is: the process through which a different constellation becomes the dominant one, shifting the functioning of the whole system. To understand societal transitions, MPA also draws upon the multi-level perspective (*ibid*) and transition pathways (Geels and Schot, 2007). The dominant constellation can adapt to the environment or landscape, or a small constellation (niche) emerges or gains power and becomes (an alternative to) the regime, or the dominant constellation adapts itself to changing societal needs.

MPA is a framework for transition stories that describes a *transition path* as a concatenation of *patterns*. Ideal-typical patterns of constellation change are: top-down reconstellation; bottom-up empowerment; or adaptive and internally induced. These patterns can be intertwined and simultaneously at play. They are driven by three possible *conditions*.

- *Tensions* that occur in the case where the environment compromises the functioning of the societal system. *Structural tension* would refer to problems with the physical, infrastructural, economic, formal and legal aspects of the relation with the environment, whereas *cultural tension* would apply to problems concerning the cognitive, discursive, normative, ideological aspects of that relation.
- *Stress,* in the case where the regime is inadequate or internally inconsistent in providing the dominant way the societal needs are met.
- *Pressure,* in the case where alternatives to the functioning of the regime emerge and become viable competitors or simply take away the need for aspects of it.

In short, transitions can be considered sequences of patterns that occur under certain conditions, producing transition paths. According to the proposers of the MPA this decoupling in patterns allows for an explanation of transition dynamics, enriching the quasi-static view of transitions by the multi-level perspective or the one-dimensional snapshot of a transition by the multi-phase concept (Rotmans et al., 2001). Ultimately, they claim, the MPA allows for greater versatility and explanatory power to better understand transitions and support further transition studies.

6.3 Multi-pattern approach applied to RftR

6.3.1 Empowerment and reconstellation pattern towards inception of RftR

Until the 1980s the FRM approach could be characterized by a 'flood defence' regime with a sectoral and technological focus, rather than an integral and spatial focus (e.g. Zevenbergen et al., 2013b; Van der Brugge et al., 2005). The constellation's structures, cultures and practices were all based on the predominant engineering solutions. FRM policies have focused on local-scale to large-scale flood protection, such as flood embankments and channelization (Saeijs, 1991). Legal, organisational and political commitments for large technology systems such as the Dutch flood defence system can lead to technical lock-in (Walker, 2000); i.e. preponderance of use of structural measures. The Netherlands is one of the most flood vulnerable countries on the planet with more than 60 % of the land area located in flood-prone areas, hosting 9 million residents and roughly 65 % of the country's gross national product (Kabat et al.

2005). Flood safety is not taken for granted, but a huge and continuous effort is required to protect the country against flooding. Currently, the Netherlands is protected from storm surges and river floods by on-going reinforcement of the flood protection system comprising coastal dunes, dikes and storm surge barriers: the Delta Works. These were developed and implemented by the first Delta Committee in the first Delta Programme in the 2nd half of the 20th century in response to the major flood disaster of 1953.

The 'flood defence' regime was challenged by *tension* from the environment. Two "wake-up" calls came in 1993 and again in 1995 when the river levels of the Meuse and Rhine almost caused dike failure, there was localised flooding and 250.000 people were evacuated. The discussions re-opened again with a fundamental reassessment of the acceptability of flood risk in the Netherlands, resulting in the outcome that the increase in economic impacts has now to be taken more specifically into account. Simultaneously, scientific evidence has grown revealing that the assumed hydraulic baseline conditions such as storm wave properties and maximum river discharges were most likely to be more severe than originally presumed and climate change and sea level rise would aggravate this (Kabat et al. 2005). An interviewed politician that has been involved with RftR from its inception stressed: "*we could not continue like this (continuously upgrading our defence system) after the 1993 and 1995 floods*". Dyke reinforcement would further encroach into the rivers' natural extent and in case of a dyke breach, potential casualties and damage would increase.

Meanwhile, alternatives to deliver flood safety in the riverine area were gaining traction and put *pressure* on the dominant constellation. There was growing societal resistance to dyke reinforcement already dating back to the 1960s. The plans of the committee for coordination of dyke reinforcement were contested because the plans required expropriation and demolition of buildings and trees. One interviewee of Rijkswaterstaat stated: "*Rijkswaterstaat was seen as Atilla on the bulldozer*". Dutch professionals presented an alternative approach in 1987 called 'Plan Ooievaar'. The plan promoted interweaving river management, nature development and landscape architecture and was followed up by several experiments. In the early 1990s the third national policy brief on water management advocated integrated water management that should consider objectives from different policy domains and for different spatial functions. Subsequently this was included in the third national policy brief on spatial planning, from which the 'NURG' programme was launched that included nature development in the riverine area. In 1992 the Dutch World Wildlife Fund published the plan Living Rivers to improve the aquatic ecosystem by introducing side channels in the flood plains that could also increase river discharge capacity. The potential of river widening measures and integrated flood risk management was explored in: Integrated exploration of Rhine branches (1992); and Exploration Space for Rhine branches (1995). These explorations provided the basis for the policy 'room for rivers' which implementation is the RftR programme (PKB, 2006). The policy stipulated the necessity to increase the river discharge capacity and storage capacity in order to ensure flood safety; complementary to merely reinforcing the dykes. River widening measures were proposed and construction in the floodplains was restricted to only river-related activities. In 2000 national parliament decided to work on the Room for the River Policy Decision (PKB) that was formally adopted in 2006 (ten Heuvelhof et al., 2007). The work on the Policy Decision led to the new 'large rivers' policy, as the old 'room for rivers' policy was considered too rigid (V&W, 2006). This new policy promoted a spatial planning and development approach and provided a new decision making framework that allowed the consideration of local and regional (spatial) policy objectives. It was enacted for the entire Dutch river system, also including the Drechtsteden and Rotterdam area. In parallel the Delta Plan Large Rivers, that was created to reinforce the dykes, tried to combine these engineering measures with nature development. In 2000, the Committee-Tielrooy promoted the introduction of a so-called 'water test' as the first legally binding action to effect integrated water management in spatial planning processes at local and regional level (CW21 2000). The year after, the Fifth Memorandum on Spatial Planning postulated water as a 'guiding principle' in spatial planning (Van der Brugge et al., 2005). A new regime emerged of integrated flood risk management combined with collaborative spatial planning, which moves away from a sectoral (water management) and regulatory (complying with protection standards) approach that focused on traditional engineering solutions to reduce the probability of flooding (Woltjer and Al, 2007; Van der Brugge et al., 2005). RftR is an iconic project in this regime change (*ibid*).

The new approach (pressure) came with structural tension, cultural tension and stress. The flood defence regime was ingrained in policy and regulation that caused *structural tension*. Even now, flood safety in the Netherlands is still regulated based on return periods of water levels from which design standards for levee systems are derived. National funding is provided only for flood defence systems and not for alternative measures (Kolen et al. 2010). The development of the PKB RftR has circumvented this as it was commissioned jointly by the Ministry of Public Works and Water Management who are responsible for flood safety and the Ministries of Housing, Spatial Planning & the Environment and of Agriculture, Nature and Fishery for whom spatial quality is their core interest in

RftR. Furthermore, the investment in river widening was justified as cost-effective to deliver flood safety based on an independent analysis conducted by the national planning bureau (Ebregt et al., 2005) prior to Government approval.

The integrated approach had to overcome *cultural tensions* because politicians and professionals were familiar with a sectoral and regulatory approach. For example, the professionals from different Ministries that presented Plan Ooievaar to the national contest 'Netherlands-Riverland', had to contribute on a personal account, rather than in name of their respective organisations that did not endorse the contribution because of its alternative approach. River widening and IFRM in general inherently require collaboration between multiple stakeholders and multiple disciplines (e.g. Potter et al., 2011). Interviewees that were involved with earlier programmes such as 'integrated exploration of Rhine branches' and 'integrated exploration lower riverine area' stated that these programmes already contributed to an increased understanding of the various interests and the need for collaboration. When in 2001 the Deputy Minister separately commissioned Rijkswaterstaat and Regional governments to develop the PKB RftR, both parties decided to develop it jointly based on positive previous experiences of collaboration?.

The structures of the old regime were internally inconsistent and not aligned with the new paradigms, which was a source of *stress*. Various authors have argued the existence of this stress. Huitema and Meijerink (2010) stated that the institutional systems surrounding water management are geared towards maintaining the <flood defence> infrastructure. Van der Brugge & Rotmans (2007) suggest that the well-organized water management regime in the Netherlands is in need of institutional renewal in terms of actors, responsibilities and tasks. New governance arrangements were implemented in RftR to overcome this stress (Rijke et al, 2012b; Van Herk et al. 2013c; in press). During the initiation phase of RftR, when the PKB was developed, regional governments could propose and design river-widening measures themselves. Interviewees mentioned especially the added value of the hydraulic model/scenario planning tool called 'box of blocks' (*blokkendoos* in Dutch) in RftR. Schut et al. (2010) explain how the instrument evolved from first a tool used by hydraulic engineers (also Reuber et al, 2005) to 'explore solution space' and calculate the hydraulic consequences of a combination of (river widening) measures, to later supporting the design and selection of measures, facilitating dialogue, cooperation and eventually decision-making between policy makers from different levels and regions. The tool was made available to all stakeholders to 'play with' and could demonstrate and visualise the effectiveness and interdependencies of measures to reduce water levels. Regional and local stakeholders could run their proposed measures through the model and nominate them for the long list of 600 measures from which the final options would be selected. The organisational structures of RftR were designed from the outset, upon adoption of the PKB RftR by national parliament (2006), to organise the relationship between national and regional authorities that was different from common practice in Dutch flood risk management. Regional governments were empowered through political agreements to deliver individual projects within the RftR programme and combine these with other spatial planning projects. A Programme Directorate was established to facilitate collaboration between regional and national government.

It is difficult to attribute this constellation change to one of the three theoretical types of transition patterns, but rather this constellation change is the result of several patterns in succession. RftR followed the *empowerment* of the small constellation that started with Plan Ooievaar that gained power and was scaled up and incorporated into national policies of the incumbent regime. The launch of the RftR programme can be considered a *reconstellation* pattern after a 'top-down' change of policy by the national Government and Parliament. Especially the deputy Minister, responsible for water management and RftR since 2002, was named in interviews as an important advocate for an integrated and collaborative approach. The initiatives and policy processes that predated RftR can be considered part of the same transition pattern. RftR contributed to the transition to IFRM by actually implementing river widening measures that combine flood safety for the Dutch river system with spatial quality. All 55 interviewees mentioned and 146 out of 151 survey respondents confirmed, RftR's contribution through its implementation. On average survey respondents rated the influence of RftR on river widening as a possible solution to increase flood safety between 'large' and 'very large' with a score of 4.24 out of 5 (standard deviation 0.82). 114 survey respondents stated they use river widening and related lessons of RftR in their own organisation for other projects. The new Delta Programme has a specific sub-programme that deals with the Dutch river system (others deal with e.g. coastal areas, the Ijssel lake area) that explicitly considers river widening as a possible solution to accommodate higher river discharges up to 18.000 m^3/s due to climate change, compared to the 16.000m^3/s that RftR aimed for. All 55 interviewees mentioned RftR contribution to improved collaboration between stakeholders, which is of pivotal importance to IFRM (Van Herk et al., 2011a). 145 survey respondents confirmed this contribution and scored 'more narrow collaboration between authorities' as 'high influence' (3.84/5). The average score is not higher and the standard deviation is 0.93, because of four negative outliers in the data. These 4 respondents are representatives of local governments that also indicated that they had not contributed significantly to RftR and also score their overall satisfaction with the programme and their respective projects much lower than average. 132 respondents use lessons from RftR for collaboration between authorities. The improved collaboration between authorities to integrate flood risk management and spatial planning was mentioned especially in interviews. 3 representatives of the Delta Programme (DP), the new large IFRM project in the Netherlands, stressed in interviews that the multi-level collaboration between authorities, as developed in RftR, is the basis for the government approach of the DP and that the DP can build upon the networks and trust created in RftR. 142 survey respondents stated the contribution of RftR to the use of a Programme Directorate as a governance arrangement to manage other large-scale programmes. 97 stressed they use a Programme Directorate in other projects, using RftR as an example. For example, the 2nd National Flood Defence Programme (HWBP2) has installed a Programme Directorate to facilitate collaboration and knowledge exchange between national and regional authorities and between projects following the example of RftR. As HWBP2 experienced cost overruns and delays, senior government officials and regional authorities urged HWBP2 to learn from RftR. HWBP2 then tried to organise collaboration between regional stakeholders and to deliver upon related (regional) policy objectives when planning and implementing measures for dyke reinforcement projects.

6.3.2 Adaptation pattern during implementation of RftR

During the implementation of RftR that started from the adoption of the PKB RftR (2006) an adaptation pattern emerged. Signs of *stress* and *cultural and structural tensions* persisted during the programme when its implementation was confronted with existing practices, methods and policy and regulation that adhered to the old regime. The inconsistencies of the new approach of RftR with the existing regime can explain the various outcomes that were mentioned by interviewees that were generated during RftR.

New methods and practices

The implementation of an alternative solution - such as river widening - needs to overcome many practical hurdles related to dominant practices (Van der Brugge & Rotmans, 2007). The RftR case study provides evidence of various outcomes in terms of practice that have been instrumental in overcoming various hurdles. Interviewees indicated that the specific hurdles were not foreseen from the start, but signalled and addressed during the implementation. Implementing river widening measures has generated outcomes in terms of new methods and practices that emerged from working together between practitioners, policy makers and scientists. They co-developed the methods and evaluated their use in practice. 95 survey respondents indicated they use new methods and tools that were developed in RftR in their other projects. 'New methods for design and analysis' was scored as having the highest influence beyond RftR (4.25) by survey respondents from the Ministry of Infrastructure and Environment (n=10) that is responsible for the new national programmes such as Delta Programme and National Flood Defence Programme. One member of staff of the Delta Programme particularly highlighted the 'box of blocks': *"we want to develop a 'box of blocks' kind-of tool to support participative planning in the Delta Programme. We will call it the Delta-Instruments and Delta-Portal that comprise: a calculation tool for water safety and fresh water supply; a method for comparison of measures; maps, descriptions and scores for measures.* Other examples include: morphologic and hydraulic modelling for groyne and quay design; physical field measurements to monitor actual water levels (comparison with hydrograph); sediments and patterns in the riverbed after flood waves (morphological); flora and fauna (ecological models); and new 3D areal measurements (laser altimetry) are being applied to study the dynamics of riverbanks. New work approaches were established and disseminated through guidelines for specific policy domains or disciplines. Room for the River has actively contributed to guidelines for: soil movement planning; planning for spatial quality; groynes information systems; consistent information requirements for hydraulic, vegetation, landscape mapping and planning; and asset monitoring and maintenance protocols. Some examples that were mentioned by interviewees are presented below.

In RftR approximately 500 groynes are being lowered along the Waal river to increase the river's discharge capacity. As a result the water level will be 10cm lower for a flow of 16,000 m3/s. Existing groynes along the Dutch rivers have been constructed and restructured in various time periods (some are more than 200 years old) for different reasons: 1. local river diversion; 2. land reclamation; 3. sediment control; and now to 4. increase river discharge capacity. The groynes are thus different in function, material and shape. RftR discovered many types of groynes in different projects and decided to create a so-called groyne-passport to document each groyne's characteristics in an open access database to support the delivery of RftR, but also for future operation, maintenance and construction of these and other groynes. Local and regional authorities, contractors and engineering firms have been invited to use and add to this information system.

Experts interviewed estimated[5] that soil excavation and movement can sum up to 30-40% of the RftR programme costs as dykes are to be relocated, flood channels are to be excavated and quays and mounds are to be constructed for assets in the outer marches. In the RftR projects that were developed first, soil experts were only actively involved in the final design stages. Compiling the lessons from these projects, the Knowledge Department of the Programme Directorate (PDR) created guidelines in 2009. Also they are actively involved in early design stages of projects that have been developed since. Cost reduction opportunities are explored to move soil within and between RftR projects, as well as with other projects. The future function and the quality of the soil are a determinant for its use elsewhere and is subject to regulation and permits. The PDR organised a parallel process to develop a programmed approach together with all directors of regional authorities granting permits and the Inspection of the Ministry. The PDR has disseminated the guidelines to external stakeholders, like engineering firms that have confirmed their use, such as the proposed soil quality map that is the legal proof required for contractors to move soil.

Much experience has been gained with designing for spatial quality. This experience is documented in RftR internal evaluations by its Knowledge department for Spatial Quality and external Quality team, and in external evaluations (Hulsker et al., 2011). Lessons are drawn from the measures and processes developed and used. The dyke is to be a leading design element as a long term investment embodying a river landscape, not to be easily compromised to local elements with a shorter life span. Also the removal of objects and local landscape structuring (e.g. zoning of functions such as nature, intensive and extensive recreational use) can increase spatial quality whilst increasing river discharge capacity, without other major interventions. Many appealing design elements were developed, taking the local land use, cultural, geomorphological and hydraulic situation and history as starting points for design elements. Process lessons relate to: the establishment of an independent Quality-team supporting local projects, the role and collaboration of landscape architects with project managers, checklist for products that anchor spatial quality in different phases such as: tendering documents, political agreements, maintenance plans, etc. The director of the Delta (sub)Programme for Rivers confirms the use of the abovementioned evaluations and has tried to involve RftR people and their approaches in the DP. The Quality team consists of external experts that use their experience when working on other projects, as will RftR employees if deployed elsewhere.

[5]more than 70% of the projects are yet to be tendered and thus costs are dependent on market bids.

Policy and regulation

Existing policy and regulation can be an important barrier to the uptake of new approaches (Van Herk et al., 2011b) and was found to have created structural tension in RftR. The interview data provided many examples of RftR's impact on policy and regulation that was inadvertently created by precedents, or deliberately managed to allow for the implementation of the Programme. During its implementation, RftR has influenced: policy for land use in outer marches; precedents for dyke requirements; regulation on soil and water quality; regulation for redevelopment of lakes; and nature-oriented planning. Some examples are described in more detail below.

When dykes are relocated inland to create more space for the river, land and assets become exposed in outer marches that were previously protected by these dykes. A special policy was developed to regulate safety standards for functions 'created' in outer marches. Based on the project 'depoldering of the Noordwaard', standards were developed for residential dwellings situated in outer marches: return period of flood damage (1/25 years) and water levels (1m water in a dwelling); availability of evacuation routes; construction requirements; hydraulic hindrance. Consequently, precedents were set to decide on the creation of earthen mounds for dwellings or expropriation, demolition and related reimbursement. For agricultural functions RftR assumed a flood return period of once every three years for the inundation of farmland. Political lobbying by agricultural organisations led to studies by the Institute for Agricultural Economics that proposed 1/25y for economically viable business that was adopted by the Dutch national parliament.

For decades dykes have mostly been renovated, but not newly designed and constructed in the Netherlands. This changed with RftR comprising of several measures including dyke relocation. Manuals for dyke design and engineering were re-interpreted and precedents were created and will be included in policy documents according to the experts at the Programme Directorate. Examples of interpretation issues that surfaced from individual projects include: the thickness of the clay-layer required on dykes; possible slopes of dykes for maintenance purposes or landscape values; and flexibility to reinforce dykes on the river-side to avoid demolition of assets on the land-side, as long as hydraulic compensation is created (to maintain the river discharge capacity).

In 2010 a new Water Act was enacted with specific elaborations on soil quality that built upon and combined the previous Surface Water and Contamination Act and Soil Quality Licence. The Act was accompanied by several guidelines such as 'Guidelines Soil' that in turn drew from Guidelines for active soil management of the rivers Rhine and Meuse. The new regulation and guidelines used a more integrated, rather than sectoral approach by regulating the sources of contamination and application of soil instead of the contaminated soil itself. This policy change enabled the movement and use of soil for a specific function that would otherwise be disapproved (e.g. dredged spoil not always needed treatment). The Knowledge department on Soil of the Programme Directorate interacted with policy makers in a parallel process, whilst facilitating individual projects. The RftR project 'Scheldener en Oldeneler buitenwaard' near Zwolle planned to fill a lake near a freshwater supply basin. The water company opposed this, which delayed the project by 6 months. The PDR commissioned research by an independent institute to prescribe requirements for landfill. The Minister for the Environment later developed policy for the redevelopment of lakes, involving researchers from the project reported here.

Nature-inclusive design, as conducted in RftR, is set as an example by the European Union's Directorate General Environment to obtain environmental objectives in flood safety and spatial planning programmes. Furthermore, RftR representatives advised the Dutch national committee, Elverding, on incorporating environmental objectives and stakeholders in the early planning stages to avoid project delays and cost-overruns in the delivery of large infrastructure programmes. The recommendations are included in a ministerial action-plan that provides guidance for such programmes. However, an interviewee stated that: ¨*Environmental regulation is not properly addressed. Often explicit wording in the regulation conflicts with flood safety measures and with the regulation's very own purpose. Nature conservation in outer marches can hinder river discharge and is not allowed and thus opportunities for dynamic nature development in these and adjacent areas are missed.*¨ Regional stakeholders sought creative solutions whilst the Programme Directorate lobbied for flexibility in the implementation of Natura2000 to national and European policy makers.

Governance arrangements and project management

During the different planning phases of RftR various new aspects of stress became apparent and project management processes have been adjusted to stimulate collaboration between the stakeholders and also the uptake of lessons learnt (Van Herk et al., 2013c; in press). The outcomes of RftR in terms of adjusted governance arrangements can be seen as an example of a contribution to an incremental institutional change such as described by Schlager (1999). RftR embraced a 'programmed approach' with 39 related projects within one programme to stimulate learning and foster a change of culture. Projects could learn from each other and create peer pressure amongst the group to stimulate progress and quality. Lessons learnt from the frontrunner projects that were ahead of other projects were transferred to the slower projects by personnel exchange, guidelines, unifying management structures and by network and training events. Various project management arrangements have been installed and optimized during the implementation of RftR that also emerged as outcomes. For the 'organisation of project teams' the survey results indicate a contribution and uptake of lessons as stated by 129 and 110 respondents respectively. 'Improved project control and risk management' was mentioned as a contribution by 136 and used in other projects by 114. Research by Albers (2012) shows the uptake of lessons in project management to deliver multiple objectives during project implementation for RftR. Objectives and integrated solutions were anchored in transfer documents between planning phases and stakeholders in RftR right up to the contractors realising the projects. For example, spatial quality was a selection criterion in tender procedures and was detailed in accompanying ambition documents.

The Programme Director appositely explained the rationale behind his decisions on governance arrangements as: "*structure should follow strategy*". The transition pattern that followed the cultural tension and stress and subsequent organisational (re)design of RftR can be classified as *adaptation*, as it was internally induced. The adaption pattern can be considered as 're-organising'. A Management Team member for the Programme Directorate explicitly highlighted this re-organisation: "*We have continuously adapted and improved the organizational structures and processes based on lessons learnt and whenever opportune. The PDR started as a local train and whilst driving we converted it into a high speed train*". Table 6.1 summarises the transition patterns for flood risk management in the Netherlands as described above with a special focus on RftR, the case study presented here.

Table 6.1. The transition path for IFRM: subsequent system states with regimes and conditions for transitional change, followed by transition patterns

System state (1953-2000)	'Flood defence' regime
Conditions	*Tension* from the environment: flood events, e.g. 1993, 1995 *Pressure*: alternatives based on IFRM, e.g. Plan Ooievaar *Structural tension:* policy & regulation based on flood defence *Cultural tension*: stakeholders used to sectoral and regulatory approach *Stress*: governance arrangements not designed for collaboration in IFRM
Transition pattern (1985-2006)	*Empowerment* of integrated approaches *Reconstellation* of IFRM by new policy
System state (2006-present)	'Integrated flood risk management' regime
Conditions	*Structural tension:* policy & regulation that hampered implementation *Cultural tension*: no practices and methods for river widening *Stress*: no detailed governance arrangements and project management structures for collaboration and learning
Transition pattern (2006-2015)	*Adaptation*: organizational re-design, practical guidelines, adjusted policies & regulation: for RftR and impact beyond
System state (present - ?)	'Adapted integrated flood risk management' regime

6.4 Discussion

6.4.1 RftR contribution to a transition to IFRM in the Netherlands

RftR is not the only demonstration of the transition to IFRM in the Netherlands, but through actual implementation this large scale programme has contributed much to the transition. A detailed analysis of the outcomes of RftR has here provided in-depth evidence of how the transition occurred in practice during the implementation of the RftR programme. The case study of this research shows that the outcomes were created to overcome barriers imposed by the old regime and to seize opportunities towards delivery of RftR within scope, time and budget. The outcomes have an impact beyond the scope and after the delivery of RftR, because of the Programme's importance for the Dutch water regime. The application of the MPA (Section 6.3) has helped to explain the outcomes that have been generated and how RftR has contributed to a transition to IFRM through these outcomes. The contribution of the RftR programme to the transition is twofold. First, the inception of a large-scale programme based on the new paradigm 'living with water' consolidated the empowerment and reconstellation of this new paradigm. Second, the actual implementation of RftR supported the adaptation of the regime to this new paradigm. The interview data has provided many examples of outcomes related to both consolidation and adaptation and the survey results have quantified the perceived or expected impact of the outcomes beyond and after RftR.

It can be observed from the case study analysis that the outcomes, the transition conditions and the transition patterns are inherently related. The tension from the environment (floods of 1993 and 1995) enabled the uptake of an alternative approach (pressure). The implementation faced stress, cultural and structural tension from the regime that was not configured for river widening as a possible solution. After empowerment of the alternative approach and reconstellation by its adoption in national policy, an adaptation pattern emerged to align structures, cultures and practices, i.e. addressing the new transition conditions. The concept of river widening as a means to increase flood safety has been of pivotal importance to the regime change. River widening required practical guidelines, new governance arrangements for collaboration and an adjustment of existing policy and regulation. The chain of patterns describes a transition path that can be considered 'teleological': a combination of empowerment and reconstellation followed by regime adaptation. Another interpretation can be that the transition path is adaptation dominated, as the regime changes autonomously in response to national policy decisions.

6.4.2 Reflection on Multi-Pattern-Approach

The case study from RftR also provides an opportunity to reflect on the Multi Pattern Approach (MPA). In the research presented here, this new method has been applied for the first time to the transition to IFRM. It has been easy to fit the framework to the case study, as others (Van der Brugge et al., 2005) already argued that the Dutch water sector has been following a transition pathway. However, the application of MPA has to some extent been ambiguous and multiple lessons can be drawn from it.

The MPA was developed for the description and understanding of the dynamics of societal transitions. The results of this research show that it can also be applied to explain the contribution of a single programme to a transition. In retrospect, the transition conditions and transition patterns explain the outcomes of RftR and, vice versa, the outcomes explain the conditions and patterns. For example, 'structural tension' (condition) from existing policy that hampered IFRM, required new policy development that will remain in force after RftR (outcome) and explains a transition pattern of 'adaptation'. The analysis of the outcomes was conducted simultaneously with the analysis of conditions and patterns and they were mutually supportive to help understand the transition path. Hence, the MPA can be enriched by analysing outcomes of various single elements (events, policies or projects) in a wider societal transition. Here, this is done for the regime level project RftR, but it should also be applicable to single micro scale experiments or macro scale landscape developments, following the multi-level perspective by Geels and Schot (2007) based on work of (Rip and Kemp, 1998). For example, the Plan Ooievaar has been analysed as a niche constellation (Van der Brugge, 2009, p. 125). Its outcomes generated pressure on the regime by proposing an alternative, integrated approach, as a response to stress generated by a sectoral approach.

As argued above, conditions, as well as outcomes have been found to be mutually interdependent in this case study. This is not necessarily a flaw in the MPA, but rather a relevant observation for those analysing and comparing transition paths. We contend that describing all transition conditions, rather than trying to combine them, supports deeper understanding of a societal transition. There is no single understanding of a transition that avoids ambiguity (De Haan and Rotmans, 2011). Section 6.3 explicitly stressed where and why the researchers have had difficulties classifying a transition condition, pattern, or path. Similarly, it has proven difficult to distinguish when a pattern begins or ends and thus to classify pattern or chain. The proposers of the MPA (ibid) argue that the choice of beginning and end point of a transition analysis are to a certain degree arbitrary and dependent on the analyst's choice. The arbitrariness on beginning and end point has shifted from pathways, when using the multi-level perspective or multi-phase concept, to shorter patterns when using the MPA. We argue that different interpretations of the timing of multiple patterns by a single or various analysts can actually enrich the analysis of a transition. The transition patterns as described in this paper are related and one could argue that they partly overlap, or neatly follow each other, or are in fact the same pattern A detailed analysis as provided in this paper that separates rather than combines the patterns was instrumental in explaining the outcomes of RftR and their contribution to the transition. The demarcation in time horizons of system states and transition patterns was influenced by the focus of this research on the case study RftR. Allowing for discussions about different interpretations of transition conditions, patterns and paths is a potential strength of the MPA. MPA gives a dynamic perspective to transitions and a higher level of detail with shorter time horizons (this research focused on a single programme that has run for a decade).

This research shows that project outcomes can be related to transition conditions and can be instrumental to a regime change. Project outcomes such as new policies, that have an impact beyond the scope of the project and are sustained after the delivery, provide a potential lever for policy makers and project managers to contribute to a transition. It was not the initial objective of this research to contribute to theories on transition management as put forward by Rotmans (2001) and enriched by many (e.g. Loorbach, 2010, Raven et al., 2010). However, the empirical data from this case study have been exploited for this purpose and some of the findings have been presented here. Loorbach (2010) provides a prescriptive governance framework in which he highlights that successful and iconic projects at the operational level can be repeated and scaled up to contribute to a transition. It is difficult to see RftR as a transition experiment at the micro-level because of its size and importance to the functioning of the flood system, but its components might. Moreover, experiments and large-scale projects can be deliberately managed alike for outcomes to address transition conditions. RftR has not adhered explicitly to objectives or models of transition management. An interviewee indicated that it was not politically viable to have an explicit objective or budget to contribute to a transition: "*Public investment assessments of the planning bureau did not include non-flood safety-related benefits of RftR, such as spatial quality, increased public support (and thus potentially reducing progress hindrance), let alone the benefit of contributing to a transition.*"

In RftR the generated outcomes were instrumental to RftR's delivery. E.g. Practical guidelines were first and foremost targeted at application within the RftR community and policy change had to enable delivery of new river widening measures as these ran into incompatible e.g. environmental, soil and other regulations. These outcomes were not foreseen beforehand to be necessary for project delivery, nor to be supportive of a transition. However, some lessons can be distilled from the case study on the governance arrangements that supported the generation of outcomes. The outcomes have arisen as a consequence from working and learning together that was stimulated from the outset by the programme's double objective of flood safety and spatial quality and the related collaborative governance approach (Rijke et al., 2012b; and Van Herk et al., 2013c). Specific governance arrangements that have contributed to learning are: cultivating a culture of learning; audits and evaluations; community building; and training programmes (*ibid*). Learning by doing is central to most adaptive management and transition management literature (an overview is presented by Armitage et al., 2008).

Also common threads in the scaling up of lessons and dissemination of outcomes were observed in RftR. Barriers and opportunities emerged at the level of individual projects and were signalled by the Programme Directorate through its project control systematics and through the stakeholder managers it had assigned to monitor and facilitate individual projects (Rijke et al, 2012b). The Programme Directorate monitored developments in terms of new policy and regulation through periodic meetings with the Ministries involved and informal contacts with parliament. Important national developments and recurrent issues in projects were scaled up to the Programme Directorate. They had the status, competences and contacts at national level to be a counterpart in discussions on national policy and regulation and to develop practical guidelines. They carefully designed shadow processes, parallel to the overall programme and project progress, to resolve these issues. Outcomes and lessons were disseminated actively to support other projects within the programme. The size of the RftR Programme, its collaborative nature, its multi-level organisation and programmed approach enabled the generation of outcomes and implicitly the management thereof.

6.4.3 Outlook to the Delta Programme and future transition patterns

Literature reviews (e.g. Zevenbergen et al., 2013b, Kabat et al., 2009) and interviews with representatives of the Delta Programme (DP), highlighted challenges for the DP programme that point to new transition conditions. The MPA has been applied tentatively here to the DP to provide an outlook to possible future transition patterns that could benefit the management of the DP and explore the value of MPA for transition management in the programme.

In 2009, the second Delta Committee was established and on November 29th 2011, the Delta Act was unanimously accepted in the Dutch Senate including the allocation of €1bn per year in the Delta Fund (Deltacommissaris2011). Following Hurricane Katrina in 2007, the Netherlands reflected on the consequences, much of which were caused by incompetent engineering and a reliance on a single approach of 'complete protection' from flooding (*tension*) (Jonkman et al. 2005). As opposed to a single approach based on current flood protection standards, the Delta Programme strives towards an appropriate balance between protection, prevention and preparedness, both now and into the future (Gersonius et al. 2010). It considers responses which have previously been outside the scope of flood risk management policies (*pressure*), such as creating more sustainable spatial urban form by accommodating excess flows that exceed the defences (in e.g. blue corridors) and minimizing damage to infrastructure and buildings. The Delta Programme has a long-term vision (up to 2100) which provides time to explore and test innovative approaches and technologies and to allow implementation over a long timeframe; thus creating the conditions under which continual improvement and adaptation can be fostered in a learning by doing process. A key element of the programme is its so-called 'adaptive delta management' (Deltacommissaris 2011). This refers to the Committee's ambition to deal with an uncertain future in a rational way by connecting long-term challenges, such as sea level rise (Katsman et al, 2011), with short-term outcomes. Adaptive delta management is a cyclical process that utilizes new knowledge to improve longer-term planning and shorter-term adaptation. Thus, longer-term plans are never complete, but are continuously adapted to changing circumstances, including those circumstances brought about by the Delta Programme's own interventions (Kabat et al. 2009). In this way, shorter-term responses that have proven successful should, where possible, be included in subsequent interventions. The approach promotes 'opportunistic adaptation' (incorporation of adaptation into urban renewal, regeneration or development and other shorter-term responses (Veerbeek et al. 2012) and 'mainstreaming adaptation' (uptake of knowledge into longer-term planning and policy processes (Gersonius, 2012).

After the installation of the Delta Commissioner in 2010, the Delta programme initiated a planning process in which so-called Delta Decisions are being developed (scheduled for 2014). Coordinated by the Delta Programme, relevant government agencies of the national, regional and local government levels collaborate with community groups and the private sector to determine and understand the challenges of the short and long term, develop potential and consecutively preferential adaptation strategies which lead to the establishment of a set of Delta Decisions that set out a plan for the future. This process can embody a *reconstellation transition pattern* that enacts adaptive delta management and thus addresses the tension and pressure described. When the DP is to be implemented after the Delta Decisions, an *adaptation pattern* can be expected, just as occurred during the implementation of RftR. Interviewees pointed to potential *structural tension* and *stress*. E.g.: "*politicians work with time horizons of 4 years, whereas the programme requires commitments to budget and plans for decades.*" Following the insights from the RftR case study, we expect outcomes to be necessary to address these conditions, but cannot foresee which. The insights on (adjusted) governance arrangements that stimulated learning and scaling up of lessons might prove valuable. Some other transition conditions have already been dealt with. The alternative integrated and collaborative approach as successfully applied in RftR has put *pressure* on the DP. The DP aims to combine flood safety with fresh water supply, but does not aim explicitly to combine flood safety objectives with spatial quality. However, as one senior staff member of DP said: "*after RftR, combining flood safety and spatial quality has become commonplace. We don't deem necessary an explicit objective for spatial quality*". The DP is organised in regional programmes to organise regional empowerment and the installation of an overarching Programme

Directorate is being considered. The (sub) Delta Programme that deals with the Dutch rivers and riverine areas builds upon the network of RftR and explicitly considers river widening as a possible solution. To use lessons learnt from RftR and to avoid potential *cultural tensions* the Ministry of Infrastructure and the Environment has proposed personnel transfer from RftR to the Delta Programme.

6.5 Conclusions

In this research the Multi-Pattern Approach (MPA) has been applied to analyse the contribution of Room for the River (RftR) to a transition to integrated flood risk management (IFRM) in the Netherlands. RftR is a 2.3 billion Euro flood safety programme that gives the rivers more room instead of merely reinforcing the defence systems. MPA is a new method to describe and understand the dynamics of societal transitions as a sequence of patterns. MPA was applied for the first time to the transition to IFRM and has provided interesting insights on the transition, whilst lessons were drawn on the use of the approach itself.

A detailed analysis of the outcomes of RftR provided in-depth evidence of how the transition occurred in practice. Outcomes that were found related to: the implementation of new river widening measures; new collaboration between authorities; new methods and practical guidelines; new policy and regulation; new governance arrangements and project management processes. These outcomes have an impact beyond the scope and after the delivery of RftR, because of the Programme's importance for the Dutch water regime. The contribution of the RftR programme to the transition to IFRM is twofold. First, the inception of a large-scale programme based on the new paradigm 'living with water' consolidated the empowerment and reconstellation of this new paradigm. Second, the actual implementation of RftR programme supported the adaptation of the regime to this new paradigm. The interview data has provided many examples of outcomes related to both consolidation and adaptation and the survey results have quantified the perceived or expected impact of the outcomes beyond and after RftR.

The application of the MPA, rather than using existing approaches such as the typology of transition pathways and the multi-level perspective, led to new insights. The MPA helped explain the constellation change as a result of several transition patterns in succession. The results of the research show that the analysis of transition conditions was instrumental in explaining the outcomes of RftR and vice versa. The separate analysis of transition patterns was used to understand how the outcomes of RftR contributed to the transition to integrated flood risk management in the Netherlands.

The MPA was developed for the description and understanding of the dynamics of societal transitions. The results of this research show that it can be applied to explain the contribution of a single programme to a transition. We hypothesise that the MPA can be applied to various individual elements (events, policies or projects) at various levels (micro, meso and macro) in a wider societal transition. The MPA can be enriched by analysing outcomes of these elements to increase understanding of the transition dynamics. The classification of transition patterns and transition paths, as well as defining their start and end point, has, however, been ambiguous and dependent on the analyst's focus on the contribution of RftR to the transition to IFRM. We contend that the ambiguity related to classifications and timings can be viewed as a strength of the MPA that enables an analyst to focus on the contribution of single programmes to a transition. We also hypothesise that transition management theories can benefit from these characteristics and can use the Dutch water sector as a case study. The Dutch water sector has moved from a sectoral to an integrated approach and is arguably shifting to an adaptive approach as advocated by the Delta Programme. A tentative application of the MPA and extrapolation of lessons from RftR provides insights for policy makers to generate outcomes and provided hypotheses for transition scientists on transition conditions and patterns they can test by monitoring the yet unknown outcomes of the Delta Programme. The outcomes will be needed to bring about a transition by overcoming new transition conditions and shaping a new chain of (adaptive) transition patterns.

Chapter 7

Adaptive co-management and network learning in the Room for the River programme

Chapter 7 Adaptive co-management and network learning in the Room for the River programme

This chapter has been published as:

Van Herk, S., Rijke, J., Zevenbergen, C., Ashley, R., Besseling, B. (in press) Adaptive co-management and network learning in the Room for the River programme. *Journal of Environmental Planning & Management.* Doi: 10.1080/09640568.2013.873364

Abstract

Adaptive co-management and learning are paramount for integrated flood management. Relevant literature focuses on adaptation at the level of physical and societal systems. The level of projects and programmes is largely overlooked, while they comprise interventions that adapt our physical systems and they provide opportunities for learning to contribute to transitions of societal systems. This paper aims to increase understanding on how learning takes place and can be stimulated within a programme. The mixed-method case study of Room for the River, a 2.3 billion Euro programme for flood risk management, shows that a programme can be organised using various governance arrangements to stimulate learning and be a means for adaptive co-management to deliver upon environmental objectives.

7.1 Introduction

Flood risk management is going through a transition from policies having a focus on flood protection to reduce the hazard probability: 'fighting against water', to actively managing flood risk that in addition to flood protection aims to reduce the exposure and vulnerability to floods: 'living with water' (e.g.: White, 2010; Newman et al., 2011; Dawson et al. 2011; Zevenbergen et al., 2013b). It is increasingly recognised that engineering responses alone cannot accommodate the future frequencies and impacts of flooding (Yovel, 2013; UNISDR, 2012; 2011a; 2011b). Multiple authors advocate a shift to integrated flood risk management (IFRM) to reduce flood risk (Zevenbergen et al., 2008; Huntjens et al., 2011; Pahl-Wostl et al., 2012). A portfolio of IFRM measures comprises in addition to hard structural measures that aim to reduce risks by modifying the system through physical and built interventions; soft structural measures that involve maintaining or restoring the natural processes with the aim of reducing risks; and non structural measures that may not require engineering, but its contribution to risk reduction is often through changing behaviour through regulation, encouragement and/or economic incentivisation (Gersonius, 2012). Land use planning is considered as one of the most crucial components in managing exposure and vulnerability to floods (Wheater and Evans, 2009; Scott et al., 2013). As physical interventions to reduce flood risk need to be incorporated in spatial planning, they need be aligned with objectives ranging as broadly as, for example, housing; nature; economics; water quality and transport, to increase the political and economic feasibility of their implementation (White, 2010; Veerbeek et al, 2012). As a consequence problems have to be framed and solutions developed for different spatial scale levels, ranging from individual dwellings to urban areas to river basins. Moreover, flood risk management has to incorporate long-term, adaptive strategies to deal with climate change and related uncertainties (Adger et al., 2005; Milly et al., 2008). Van Herk et al. (2011a; 2013a; 2013c) stress the governance challenges this poses for the implementation processes of IFRM policies and investment projects. They require: collaboration between multiple stakeholders; combination of objectives and funding from various policy domains; consideration of a wide range of possible solutions at multiple spatial scale levels (from local, to international) and for different time horizons (short-, medium- and long term); and the involvement of multiple disciplines. Collaboration and learning[6] on the part of the stakeholders involved is needed to deal with these governance challenges: to implement IFRM policies

[6] "Learning is taken to be a process of long-lasting change in the behaviour or the general ability to behave in a certain way, which is founded on changes in knowledge and beliefs" (Siebenhüner, 2005).

and change the physical system (natural and man-made); and to stimulate a transition of the societal system[7] to IFRM (e.g. Pahl-Wostl, 2009).

An increasing amount of literature in environmental management and integrated water management has emerged introducing the concepts of co-management, adaptive management, adaptive co-management and adaptive governance (e.g. Huitema et al., 2009; Armitage et al., 2008). The concepts address many elements related to the governance challenges of IFRM. They have various and partly overlapping definitions that comprise dimensions of learning and collaboration (*ibid*, Plummer and FitzGibbon, 2004; Berkes, 2009) to adapt to the complexity, uncertainty (e.g. due to climate change) and dynamics of the physical systems (e.g. Folke et al., 2005; Huitema et al., 2009). Various authors on adaptive co-management and adaptive governance (Ruitenbeek and Cartier, 2001; Olsson et al., 2006; Folke et al., 2005; Dietz et al, 2003) highlight the need for flexible structures for multi-stakeholder management and learning to deal with the multi-objective reality and the dynamics of physical systems. However, there is limited practical guidance and empirical evidence on the role of learning in adaptive co-management. As Armitage et al. (2008, p87) ask: "if learning is acknowledged as being of central importance to adaptive co-management and related governance models, why is it usually employed in an everyday, familiar sense with little detailed examination?".

Learning is also needed to stimulate a transition of the societal system to IFRM to enable and facilitate a widespread uptake and delivery of IFRM (e.g. Pahl-Wostl, 2007; 2009; Van der Brugge and Rotmans, 2007; Van Herk et al., 2011b). Transition literature defines a transition as a fundamental change in the structures (the formal, physical, legal and economic aspects of functioning restricting and enabling practices), cultures (the cognitive, discursive, normative and ideological aspects of functioning involved in sense-making of practise) and practices (the routines, habits, formalisms, procedures and protocols by which actors, which can be individuals, organisations, companies, etc., maintain the functioning of the system) of a societal system, profoundly altering the way it functions (e.g. Rotmans and Loorbach, 2006; Rotmans and Loorbach, 2009; Van Raak, 2010). Transition theories (e.g. Rotmans and Loorbach, 2006; De Haan and Rotmans, 2011) help describe and understand the dynamics of societal transitions on the level of the

[7] A societal system is a part of society that can be attributed a functioning and functioning is the way a societal system meets a societal need. The functioning of societal systems can be described by its: structures, cultures and practices. (e.g. Rotmans and Loorbach, 2006; Rotmans and Loorbach, 2009; Van Raak, 2010).

regime[8], but do not focus on the role of projects or programmes during a transition. Van Herk et al. (2011b; 2013b) show that the actual implementation of a project or programme can support the adaptation of the societal system as it also generates outcomes that have an impact beyond the scope of the project and that are sustained after the delivery of the project. Outcomes arose in the case studies they examined as a consequence from working and learning together, but the learning processes and learning outcomes were not studied (*ibid*).

Two bodies of literature, inter alia, adaptive co-management (ACM) and transition management (TM) provide learning-oriented management frameworks that stress the importance of collaboration and participation of stakeholders that are involved in or affected by the management and change of the physical and societal systems. Various authors have developed conceptual frameworks to classify the goals, outcomes and approaches to various types of learning by individuals, organisations and networks (e.g. Armitage et al., 2008; Steyaert and Jiggins, 2007; Pahl-Wostl, 2009; Folke et al, 2005; Huitema et al., 2009). They focus either on learning to adapt the physical system (ACM), or on learning to adapt the societal systems (TM). The two bodies of literature also use various other, overlapping terminologies, such as social-ecological system (the co-evolutionary units of social and ecological systems [Folke et al, 2005])and socio-technical system (all the physical systems, actors and rules required in order to perform a particular function [Geels, 2005]). These terminologies combine 'ecological' or 'technical' with 'social' to indicate the powerful reciprocal feedback loops between the systems. Hence, the attempts of abovementioned authors to develop (learning) theories that comprise both the adaptation of the physical and societal system. However, the bodies of literature provide no guidance or an evaluation framework for learning on the level of a project or programme, but rather conceptualised theories on the system level. Contrary, this paper starts with the hypothesis that learning for both system changes occurs simultaneously in IFRM projects. The research presented in this paper has taken an individual programme as the level of analysis. A programme can be defined as "a group of related projects managed in a coordinated way to obtain benefits and control not available from managing them individually" (PMI, 2008, p.434). A project is defined here as a temporary endeavour with a defined beginning and end (usually time-constrained, and often constrained by funding or deliverables, undertaken to meet unique goals and objectives, typically to bring about beneficial change or added value (based on: Nokes, 2007).

---

[8] The societal subsystem that dominates the functioning of the system (De Haan and Rotmans, 2011)

The level of projects and programmes is relevant for multiple reasons. The proactive adaptation of the physical system takes place through interventions that are part of various projects or programmes, such as the one studied for this research, that together form adaptation pathways (Gersonius, 2012). Moreover, programmes present opportunities for learning-by-doing (e.g. Armitage et al, 2008) and opportunities to work together that enable effective learning (Jentoft, 2007). Effective co-management requires flexible, multi-level governance systems designed to enhance institutional interaction and experimentation to generate learning, but there is little experience on how to accomplish this (Berkes, 2009). Institutional experimentation and learning can serve as a prelude to finding the right mix of self-governance, co-governance, and hierarchical governance specific to a situation (Kooiman et al., 2005). Programmes can provide flexibility for learning and experimentation, because they are designed for a specific situation and purpose rather than forming part of the formal institutional framework (Rijke et al., 2012a). Shehu and Akintoye (2009) recognised the importance of incremental programme design and adaptation to changing contexts. Because programme management contexts are complex, programmes should be organised as complex adaptive systems (Ritson et al., 2011). Understanding the learning within programmes is paramount, because programmes are increasingly being adopted to implement organisational transformational strategies and integrate multiple projects (Maylor et al., 2006). Finally, programmes that stimulate learning provide an opportunity to obtain empirical data and evaluate learning. Armitage et al. (2008, p97) concluded that: ¨Attention to learning as an explicit strategy in the design and operation of co-management is only just emerging. There is little experience upon which to base the development of best practices, or critically assess the process of learning in adaptive co-management. Thus, an important task is to identify and consistently evaluate those instances where learning (as a learning-by-doing process or through 'controlled' adaptive experimentation) is an explicit concern, to identify what works and what does not, and to elucidate key lessons and helpful models for future governance innovation.¨

This paper aims to increase understanding on how learning takes place and can be stimulated within a programme. This research analyses a case study on (i) the learning outcomes of a programme that contribute to adaptation of the physical and societal system or governance arrangements[9]; and (ii) how governance

[9] Governance arrangements are defined here as the organisation of processes and structures required for steering and managing parts of societies (Kooiman, 1993; Pierre and Peters, 2000); here within programmes. As a process, a governance arrangement refers to the organisation to manage networks and the collaboration

arrangements within a programme have stimulated that learning. Theoretical notions from ACM, TM and social learning are applied to a single case study. As the ACM, TM and social learning literatures provide no unambiguous way to analyse learning outcomes of a programme, this paper also uses theory on network learning from a different domain, namely health services (Knight and Pye, 2004). The selected case study is Room for the River (RftR). RftR is a 2.3 billion Euro flood protection programme in the Netherlands that comprises 39 river-widening measures to allow for higher river discharges and to improve the spatial quality along the Dutch river system. RftR is a unique programme that adopts an adaptive co-management approach to IFRM, as a deliberate alternative to a technocratic engineering-focused approach that has been commonplace to date in the Netherlands (Zevenbergen et al., 2013b). Its approach to IFRM and governance is considered exemplary in a national and international context (Kabat et al., 2009). The innovative approaches of Room for the River required social and network learning to deliver its flood risk, environmental and wider societal objectives. The analysis as presented in this paper, of the RftR's learning outcomes and governance arrangements that stimulated learning, can support the design of future programmes. The perspective of a programme can enrich learning; adaptive co-management; and transition management theories that to date mostly focus on societal and physical systems as a whole. Future research will be necessary to validate the wider applicability of findings, as the scope of the research presented in this paper was limited to merely a single case study.

7.2 Network learning outcomes in a programme

It is not straightforward to analyse learning outcomes in a programme characterised by adaptive co-management. Armitage et al. (2008) have given an overview of different definitions and typologies of learning, such as experiential learning; transformative learning; and social learning. They state that even within a particular framework there are multiple and sometimes contradictory definitions of learning. In this paper, we elaborate on social learning because of its explicit focus on the underlying multi-stakeholder participatory process that characterises policy programmes in integrated water management (Huitema et al., 2009; Pahl-

---

between individuals and organisations in a programme (adapted from: Kjær, 2004; Rhodes, 1996). A governance arrangement as a structure refers to the design of patterns and mechanisms in which social order is generated and reproduced within the programme (adapted from: Voß, 2007).

Wostl et al., 2007; Blackmore et al., 2007). Armitage et al. (2008) define social learning as a process of iterative reflection that occurs when we share our experiences, ideas and environments with others. Muro and Jeffrey (2008) have reviewed studies of social learning in participatory processes. They compare different definitions and conclude that ¨social learning requires the communication and interaction of different actors in a participatory setting which is believed to result in a set of social outcomes, such as the generation of new knowledge, the acquisition of technical an social skills as well as the development of trust and relationships which in turn may form the basis for a common understanding of the system or problem at hand, agreement and collective actions.¨ Most authors of social learning adhere to the same learning typology that distinguishes: single loop learning; double loop learning; and triple loop learning. These concepts have been developed by King and Jiggings (2002), Hargrove (2002) and Keen et al. (2005) and have been adopted and applied extensively in a variety of ways.

Table 7.1. Different interpretations for learning typologies: single, double and triple loop learning

	Huntjens et al.,2011; Pahl-Wostl, 2009	Flood & Romm (1996)	Tuinstra (2008)	Farrely & Brown (2011)	Armitage et al. (2008)
Single loop learning	Refinement of established actions	Do things right	Instrumental learning adopting new knowledge to existing frameworks of objectives and causal beliefs	Technical learning to achieve objectives	Change actions and strategies
Double loop learning	Changing guiding assumptions / reframing	Do the right things	Change beliefs, norms and objectives	Conceptual learning that reconsiders objectives	Change values and policies
Triple loop learning	Regime transformation or paradigm shift in the structural context.	Power imposing values and norms or vice versa	Learning the ability to learn itself.	Social learning for transformation from technical to conceptual	Change governance norms and protocols that predicate single and double loop learning

Table 7.1 illustrates what Muro and Jeffrey (2008) state as a lack of a consistent concept of social learning that complicates the task of defining common indicators to measure social learning as either a process or outcome. Moreover, none of the literature listed in Table 7.1 explicitly includes learning in individual programmes, or the learning that supports the adjustments of governance arrangements herein, but rather the learning in a sector (e.g. flood management) as a whole. Huntjens et al. (2011) and Pahl-Wostl (2009) classify the learning that contributes to changes of the regime as triple loop learning, referring to structural changes in the institutional context. Pahl-Wostl (2009) assumes that the learning in formal policy cycles, to which individual programmes belong, is restricted to single-loop learning. However, Van Herk et al. (2013c) showed that a programme adopting new paradigms of IFRM and new approaches for adaptive co-management will also foster double and triple loop learning. Such programmes can be considered large-scale experiments (Cook et al, 2004), even though they are not implemented with the aim to experiment or learn, but rather learning is instrumental to deliver specific policy objectives. All authors from Table 7.1 focus single loop learning on changing actions, or technical learning within existing frameworks, and not on changing governance arrangements within programmes or projects. The refinement or change of actions (single loop) and frames, values or beliefs (double loop) is also not positioned or evaluated within the framework of an individual policy or investment programme that delivers adaptation of physical or societal systems.

As the social learning literatures provide no unambiguous way to analyse learning outcomes of a programme, this paper uses theory on network learning from a different domain, namely health services (Knight and Pye, 2004). Literature on network learning – learning by a group of organizations as a group (Knight, 2002) – partly overlaps, but is also complementary to social learning theories. Policy and investment programmes in IFRM can be considered networks. A 'whole network' consists of multiple organisations linked through multilateral ties (Provan et al., 2007). IFRM involves interlinked organisations ranging from river basin authorities, to national governments, regional and local governments for spatial planning and urban water systems. Networks are often formally established and governed and goal directed rather than occurring serendipitously (Kilduff et al., 2003). Water management networks comprise multiple organisations with formalised roles and political mandates to manage amongst others: water quality; flood protection and drought. Knight and Pye (2004) analyse network learning as the change in network level properties, such as density and connectedness of organisations, structures and centralization in networks, or governance among a range of organisations. Based on a case study they classify learning outcomes into

3 categories: interpretations; structures; and practices. Changes in 'interpretations' overlap with double loop learning and changes in 'practices' overlap with single loop learning. Changes in 'structures' partly covers triple loop learning, if referring to changes to institutional structures, but adds a new perspective in comprising changes to governance structures in a single programme. This research applies the classification of learning outcomes by Knight and Pye (2004) to integrated water management and defines its components as follows:

- Learning in terms of changing *interpretations* refers to the dominant philosophies or paradigms. In the scope of this research they are related to flood risk management, spatial planning, programme management or multi-level governance[10].

- Learning outcomes in terms of changing *structures* are the patterns that are being (re)designed or that emerge from governing activities of social, political and administrative actors (Kooiman, 2003). These patterns also comprise the governance arrangements to manage a programme or project, in order to organise networks of actors and institutional frameworks.

- Learning in terms of *practices* comprises cognitive and behavioural learning related to working together in a programme or project on flood risk management and spatial planning.

7.3 Research approach

In exploring how learning in a programme that aims to adapt the physical system takes place, the research approach presented here aimed at analysing: (i) the learning outcomes of a programme; (ii) how governance arrangements within a programme have stimulated that learning; and (iii) how learning supports the adjustments of governance arrangements within a programme. Room for the River (RftR) has been selected as a case study for two reasons. Firstly, it has adopted an adaptive co-management approach to integrated water management and is a crucial step in the transition to 'living with water' (Van der Brugge et al, 2005; Van Herk et al., 2012b; 2013c). Secondly, Rijke et al. (2012b) show that the programme has an exceptionally high performance in terms of project output, stakeholder satisfaction, budget and time (Rijke et al., 2012b). Learning and programme adaptation have been identified as success factors (Rijke et al., in

[10] The outcome of interaction among multiple actors from different sectors with different levels of authority (multi-level governance; Agrawal (2003)

press). RftR has a long duration (1999-2015) that allows for much learning and requires adaptation to deal with ever changing internal and external complexities such as: changing stakeholder interests and configurations; new scientific insights; political and economic developments (Hertogh et al., 2008; Hertogh and Westerveld, 2010).

A combination of document analysis; face-to-face interviews (n=55); and observation at 3 training events (45 participants) and 2 political conferences (approximately 220 participants) have been used to analyse the learning outcomes in terms of interpretations, structure, practices (Section 7.2). First, we established the baseline for the learning outcomes at the start of the programme based on an extensive document analysis of formal policy and programme documents. Interviews with individuals who were involved with the initiation phase of the programme (n=10) were used to verify that baseline and further clarify the selection of governance arrangements or structures and practices to work with the new paradigms of river widening and co-management that were set at the start of the programme. The interviews were semi-structured and covered: the motivation for the programme objectives and river widening concept; the design of the RftR programme organisation and of activities for multi-actor collaboration and public participation. Section 7.4.1 describes the learning outcomes for the initiation phase of RftR.

Further semi-structured interviews were conducted to explore learning outcomes as changes in interpretations, structures and practices during subsequent stages of the programme. Questions were asked as to which changes occurred and what induced these changes. 31 interviewees were directly involved in the design and realisation stages of the program and 14 other interviewees held strategic positions at the levels of senior policy makers and decision makers. Interviewees represented a range of disciplines and organisations involved with the individual projects (i.e. waterboards, provinces, municipalities and Rijkswaterstaat, the executive arm of the Ministry of Infrastructure and the Environment that is responsible for the design, construction, management and maintenance of the main infrastructure facilities in the Netherlands) and the programme as a whole (i.e. the Room for the River programme Directorate, its mother organisation Rijkswaterstaat and the Ministry of Infrastructure and Environment). Interview transcripts were analysed using QSR Nvivo 9. Data was coded following the interpretations, structures and practices as found in the design phase. New codes have been added when new categories of learning outcomes emerged from the data. Observed changes have been verified through additional document analysis of meeting notes at all levels of programme governance and of periodic update reports of project teams to the Programme Directorate and from the Directorate to Dutch parliament. Section 7.4.2 presents the learning outcomes for the design phase of RftR.

Recurrently mentioned and similar learning outcomes and changes to interpretations, structures and practices have been clustered and analysed for interdependencies. We found mutual dependencies between the 3 types of learning outcomes and their evolution during the various stages of the programme. In Section 7.4 the learning outcomes have been presented and explained as clusters. This paper focuses on the learning outcomes at the network or programme level. For more learning outcomes, see Van Herk et al. (2013b) where the lessons from RftR that have been taken up by individuals, organisations and future programmes have been analysed and considered in terms of their contribution to a transition to IFRM.

Finally, we have analysed the interview data for governance arrangements that have stimulated learning and compared these with literature on learning strategies in adaptive co-management (e.g. Cook et al, 2004; Berkes, 2009; Huitema et al, 2009). From the interview data, we have inductively devised codes for QSR Nvivo 9 on these governance arrangements. These codes largely coincide with the institutional prescriptions for adaptive co-management of Huitema et al. (2009) and strategies to facilitate or improve co-management of Berkes (2009). In Section 7.5 we present the learning mechanisms that took place in RftR following the abovementioned theoretical ideas. This research provides empirical evidence and operational lessons as to how these theoretical prescriptions have been applied in a programme. A quantitative survey (n=151) has also been used to understand the importance of governance arrangements in terms of programme output and its contribution to a wider transition. Respondents were asked to rate success factors of RftR for its output and uptake in future programmes, using a five-point Likert-based scale, where 1. is very unimportant and 5. is very important. These success factors were previously deduced from the interviews and comprised: the urgency after the floods of 1993 and 1995; the programme objectives; its organisation; human factors such as leadership and trust; contextual factors such as political and economical developments. These results cannot provide evidence of a causal relationship between governance arrangements and learning and its impact on output and outcomes because: multiple factors influence the impact; governance arrangements have been continuously adjusted; and learning cannot be qualified and measured (Muro and Jeffrey, 2008). Nonetheless, the results provide insight into their relative importance of governance arrangements compared to other succes factors with e.g. the impact of the floods of 1993 and 1995 that scored an average of 4.3 out of 5. To validate the findings, we organised a network event to discuss the lessons learnt from RftR (approximately 150 participants that have been involved in the programme in various roles). The research at the level of a programme required the use of multiple sources of evidence and a validation from multiple perspectives.

7.4 Learning outcomes in Room for the River

7.4.1 Initiation phase 1999-2006, learning from previous projects and align learning outcomes with context

The floods of 1993 and 1995 in the Netherlands together with increased attention to climate change have contributed to a paradigm shift in flood management towards accommodating floods, i.e a change of *interpretations*. There has been growing societal resistance to dyke reinforcement and a growing belief that these measures alone could not deliver future flood safety (Van der Brugge et al., 2005). The projects Plan Ooievaar (1986) and the Meusse Works (1997) were already exploring the possibilities of river widening measures to reduce flood risk and to deliver on other policy objectives thanks to the stronger spatial planning component of river widening than traditional measures such as dyke reinforcement alone (*ibid*). RftR was launched to increase flood safety by giving the rivers more room, combined with increased spatial quality of landscape, nature and culture (Schut et al., 2010; Van Herk et al., 2012a; Rijke et al, 2012b). Survey analysis from the work presented here has shown that this double objective has been one of the most important success factors in the realisation of the programme (scoring 4.03/5).

New governance arrangements or 'structures' were deemed necessary to deliver river widening and reach both objectives. Firstly, the programme directors decided to commission an evaluation of the programme objectives to two independent, renowned bodies in order to increase credibility and accountability. In previous large-scale infrastructure projects such as the Betuwe Rail Freight Route the project objectives where continuously questioned (Hertogh et al., 2008). Deltares, a research institute, undertook the flood modelling for the entire river basin and evaluated the potential decrease of water levels from river widening measures. An independent expert panel on spatial quality was created, the Quality-team or Q-team, to evaluate and safeguard the programme's objective of spatial quality. Secondly, river widening and setting the explicit objective of spatial quality required the involvement and active participation of regional stakeholders with interests and competences in spatial planning. Regional and local stakeholders could propose measures themselves and the selection of 39 measures out of a long list of 600 was done by two committees comprising political representatives of regional and local authorities.

This participative planning approach had crucial consequences for learning outcomes in terms of 'practices' as it enabled social learning, i.e. 'working and learning together' (Pahl-Wostl et al., 2007). Participatory planning was supported by a computerized hydraulic model/scenario planning tool known as 'box of blocks'. Schut et al. (2010) explain how the instrument evolved from an initial tool used by hydraulic engineers (Reuber et al, 2005), to 'explore solution space' and calculate the hydraulic consequences of a combination of river widening measures, to later supporting the design and selection of measures, facilitating dialogue, cooperation and eventually decision-making between policy makers from different levels, jurisdictions and regions. Stakeholders learnt what type of measures could reduce water levels and what other amenities these measures could offer. Regional stakeholders came to see opportunities to combine nature, recreational, industrial or urban development. Their enthusiasm for these associated policy objectives reinforced the idea behind the programme objective of 'spatial quality'. According to interviewees from the National Ministry: ¨the box of blocks was a crucial element for effective collaboration between authorities¨.

Changed 'interpretations' towards programme management also promoted regional leadership and related governance arrangements. The previous large scale infrastructure projects High Speed Rail Line and Betuwe Rail Freight Route had significant cost overruns and time delays because of regional opposition (Hertogh et al., 2008). These experiences led to a new view on participatory planning with regional stakeholders. The Ministry of Infrastructure and Environment and Rijkswaterstaat, their executive arm, had also commissioned and managed these previous projects and were of the view that a new collaborative approach was necessary for successful delivery of RftR. At the start of the design phase of RftR, the Dutch Government commissioned the planning and delivery of the 39 individual measures mostly to local and regional authorities. The national commission Duivesteijn that had evaluated the High Speed Rail Line and Betuwe Rail Freight Route in 2004, concluded that project control had not been strict enough and that more accountability was necessary (Commissie Duivesteijn, 2004). The national government felt ¨it cannot go wrong again this time¨ and wanted more insight and control on project progress. Room for the River became a so-called 'Large Project' for the Dutch Parliament; a newly created status that required 6 monthly updates to Parliament to improve transparency and accountability. The Dutch Government assigned the implementation of the Programme to the Programme Directorate Room for the River (PDR) that would serve as an interface between national and regional governments and could safeguard strict programme controlling and management. Regional authorities justified progress on their project to the PDR. In turn, the PDR had to justify progress of the Programme every 6 months to Dutch Government and Parliament for which it monitored the progress, scope and quality of the 39 projects. The underlying management philosophy was 'controlled trust': ¨We (the Dutch government) trust you (regional authorities) will deliver, but we will control you (via the PDR)¨. Important lessons on management structures were learnt from the Meuse Works, a river widening project that predated RftR. Its monitoring system was furthered and organised around decision-making milestones set by the Programme Directorate. Also, the Programme Directorate was deliberately staffed with professionals with experience from the previous Large Public Works projects. Figure 7.1 summarises the learning outcomes in interpretations, structures and practices during the initiation phase and the input from previous programmes.

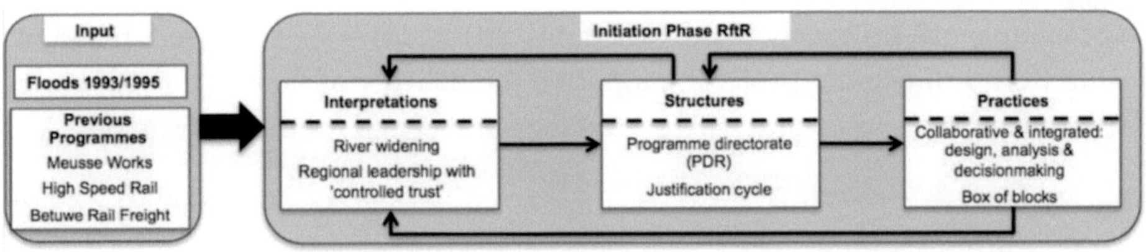

Figure 7.1. The learning outcomes in interpretations, structures and practices during the Initiation phase, and the Input from previous projects

7.4.2 Design phase 2006-2011, adjusting interpretations, structures and practices.

A Management Team member of PDR explicitly highlighted the learning during RftR: ¨We have continuously adapted and improved the organizational structures and processes based on lessons learnt and whenever opportune. The PDR started as a local train and whilst driving we converted it into a high speed train¨.

The advantages and pitfalls of a 'programmed approach': 39 related projects within one programme became apparent to programme managers during the design phase. The 'interpretation' emerged that the programme's success depended on the success of its' individual projects. As one programme manager put it: "RftR is like a chain of pearls: beautiful individual projects that cannot be considered in isolation of each other." Each individual measure is necessary to reach the overall objective of increasing river discharge capacity and in lowering water levels along the river basin. Also, the satisfaction or dissatisfaction amongst stakeholders in an individual project could affect other projects. As a consequence the PDR emphasised to regional stakeholders the importance of their project in achieving the overall programme objectives and thus the flood safety of the Netherlands. Moreover, programme managers saw the potential that projects could learn from each other and create peer pressure amongst the group to stimulate progress and quality. These insights were exploited through 'structures' by explicitly naming frontrunner projects that were ahead of other projects in terms of planning. Lessons learnt from the leading projects were transferred to the slower projects by personnel exchange, guidelines, unifying management structures and network and training events. As an example, the project team for the 'Noordwaard' project, a frontrunner that had already entered the construction phase, started also working on another project 'Ruimte voor de Lek' and some team members started working part-time for yet another project in 'Lent'. Guidelines were developed for: soil movement planning; planning for spatial quality; groyne information systems; consistent information requirements for hydraulics, vegetation, landscape mapping and planning; and asset monitoring and maintenance protocols (Van Herk et al., 2013c; 2013b). They were based on lessons from frontrunner projects, supported the delivery of slower projects within RftR and are being used and will be used for parallel and future programmes.

The management structures were established uniformly amongst the 39 project teams to further stimulate the exchange of lessons between projects and their team members who had similar roles. Permit request coordination teams were established in all projects to coordinate and combine permit requests to the various authorities, having observed that multiple permit procedures could hamper the progress of projects. A variety of authorities are responsible for granting permits such as: nature permits, building permit, transport permits, etc. Politicians from the steering committees and civil servants from the project teams tried to streamline permit requests amongst different authorities and interacted with their independent permit departments. This was beneficial as all permits were then obtained together, instead of going through multiple formal procedures. This was to ensure that conflicting interests did not lead to a permit being granted on one aspect, but being rejected on another, possibly delaying the whole project. The interaction in the working groups increased understanding of mutual interests and supported the search for integrated solutions. The same learning approach of transfer and replication of lessons in subsequent projects has been applied to the individual projects. For example, a project that comprised the lowering of approximately 500 groynes along the Waal river was organised into 3 sequential tranches so as to learn from previous tranches in terms of: hydraulic;

morphologic and ecologic effects; construction time; market approach. 'Practices' for groyne designs were adapted after the first tranche, based on stakeholder feedback and new insights as to the effectiveness of the designs'.

The paradigm of 'controlled trust', no central government interference in individual projects championed by regional authorities, was adjusted during the design phase of RftR. The PDR observed from its monitoring activities that the progress and quality of the 39 projects often failed often due to similar issues. The PDR changed its 'interpretations' and decided to pro-actively support projects: i.e. the monitoring led to a culture of collaboration as the PDR facilitated and supported the individual projects to comply with the criteria that were monitored. Organisational 'structures' governing the interface between PDR and regional projects were changed to embed the facilitation activity. Project managers contacted the Programme via stakeholder managers at the PDR that monitored the regional political support. However, many challenges that were encountered in projects were not political, but related to specific technical and project controlling issues. The PDR decided to reinforce the front-office (i.e. stakeholder managers), with so-called triangular meetings with representatives from 3 departments: stakeholder management; knowledge (eg soil quality, piping and cabling, hydraulics); and project controlling (budgeting, contracting). The PDR introduced a requirement for progress and risk management reports with criteria from these 3 departments. Where there were recurring issues covering several projects, these were addressed by specific task forces that in certain cases even managed to change national policies where they prevented progress, such as: land use in outer marches; precedents for dyke requirements; regulations on soil and water quality; regulation for redevelopment of lakes; and nature-oriented planning. This benefited all RftR projects and future programmes (Van Herk et al., 2013b). Also, the Q-team became more instrumental in supporting individual projects. The Q-team not only evaluated, but also supported and promoted attractive spatial designs and evaluated an individual project's technical feasibility. They periodically visited projects and provided independent advice on project designs and the design process. The monitoring activities of the PDR have been converted into collaborative activities with regional stakeholders: 'collegial monitoring'. Regional stakeholders requested permission to carry out or commission the monitoring activities themselves to increase their own learning experience.

The interpretations of the spatial quality objective changed over time. Instead of presenting river widening as an opportunity for delivering spatial quality and to incentivise regional stakeholders, this objective is now being seen as a luxury. The political landscape changed significantly after the 2010 national and local elections. Political support for spatial quality decreased, in particular budgets for nature development and recreational development, whilst agriculture as economic activity gained relevance. Representatives of the Programme Directorate explicitly highlighted that a programme with a long duration needs to be able to adapt to changing context and be flexible. A respondent responsible for communication explained that the programme no longer used the word 'nature' in communication and focused on flood safety and agriculture. Moreover, solutions were sought to preserve or develop farmland. E.g. in the Veessen-Wapenveld project where a by-pass is proposed through farmland, stakeholders decided to acquire additional agricultural land outside the area to compensate nature development in the river IJssel by-pass (Van Herk et al, 2012b). Figure 7.2 summarises the learning outcomes in interpretations, structures and practices and their interdependencies during the Design phase.

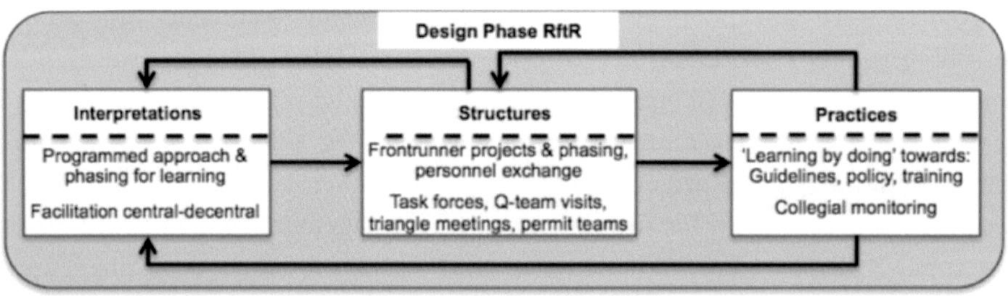

Figure 7.2. The learning outcomes and their interdependencies during the Design phase

7.5 Governance arrangements for learning and adaptive co-management in RftR

Several authors provide guidance on strategies or governance arrangements for learning based on extensive reviews of studies into adaptive co-management (e.g. Cook et al, 2004; Huitema et al, 2009, Berkes, 2009). Cook et al. (2004) present a list of experimental approaches that can foster learning in human-environment interaction. Huitema et al. (2009) devise 4 institutional prescriptions for adaptive co-management. Berkes (2009) lists strategies that have been used to facilitate or improve co-management. The strategies they advocate partly overlap and do not focus on individual programmes, but rather on physical and societal systems. However, we have found empirical evidence of how their strategies have been applied in the RftR programme.

Huitema et al. (2009) propose 'polycentric governance' with dispersed centres of power that tap into capacities at different geographical scopes. Polycentric governance is inherent in integrated water management and at the core of the governance arrangements of RftR. The 39 individual projects are championed by regional authorities and monitored by the Programme Directorate (PDR) that operates at national level. Berkes (2009) refers to 'fair and democratic distribution of power' observed in RftR by regional stakeholders getting the design freedom to (learn to) devise measures that combine flood safety with spatial quality. The regional leadership required a justification cycle to monitor progress and quality of projects by the PDR. Dutch parliament requested progress updates and the Ministry commissioned independent audits by scientists and consultants. These are expressions in RftR of 'downward accountability' and 'collaborative monitoring' (*ibid*) that fostered learning. In general RftR fostered a culture of transparency and accountability that was operationalized in a wide range of audits, evaluations and monitoring assessments (Rijke et al., 2012b) from which many lessons were drawn. Technical and political hurdles have been observed and could be overcome jointly. For this, the PDR was instrumental in 'bridging knowledge' (*ibid*) in mobilising knowledge, skills and capacities of actors at different levels. The governance arrangement of central boundaries, with decentralised leadership was considered an important success factor in RftR (3.84/5) and survey respondents scored the importance of clear objectives and strict project management at 3.99/5. The effect of these governance arrangements extended beyond the programme. Room for the river has had a strong influence on greater collaboration between different authorities in water management: 3.84/5. As such RftR has provided a positive example for the national commission Elverding, that had urged managers for large infrastructure projects in the Netherlands in 2008 to apply improved planning processes for faster and better results; especially through earlier participation of stakeholders (Commissie Elverding, 2008). The survey results presented here also indicate that RftR has had a major influence on the (future) application of programme directorates (3.86/5). A senior government official confirmed that: ¨the Delta-Programme is currently considering the creation of Programme Directorates, following the example of Room for the River¨.

Huitema et al. (2009) propose 'public participation' to support decision making amongst multiple stakeholders and increase democratic legitimacy of the decision-making process. Berkes (2009) presents various facets of public participation that stimulate learning that have been observed in RftR. The box of blocks that was used during the initiation phase of RftR to devise and select river widening measures is an example of 'participatory scenario building' (*ibid*). Interviewees have stated their increased insight in: the functioning of the river system; in the mutual dependence of measures to reduce water levels in the river; and hence in the need for a river-basin approach or 'watershed approach' (Cook et al, 2004) or need for a 'bioregional perspective' (Huitema et al., 2009). 'Co-production of knowledge' and 'participatory research' (Berkes, 2009) have been observed mostly in the 39 individual projects. Van Herk et al. (2012a) describe the social learning processes in RftR that combine: system analysis, collaborative planning, design and engineering; and governance, to implement the individual measures. Local farmers and inhabitants have contributed to the analysis by providing local knowledge. In the project Overdiepse Polder they have provided the design concept for the spatial plan. Another example is the joint collegial monitoring by regional and national authorities that has been stated as a learning experience. Berkes (2009) also lists 'cooperation building tactics' that have been observed in RftR through: stakeholder managers of the PDR that foster regional political cooperation; the Q-team regularly visiting projects to advise on design work to increase spatial quality considering ecology, hydrology: morphology and landscape architecture; personnel exchange between projects.

Huitema et al. (2009) propose 'experimentation' as a form of trial and error learning and Cook et al. (2004) evaluate a range of approaches for experimentation. Learning at a programme level can be considered 'Adaptive Management' (*ibid*) as it has been a 'learning by doing' process, at least partly 'directed at policy and management modification'. As such RftR can be considered as one 'large-plot experiment' (*ibid*). The governance arrangement that was mentioned most by interviewees to stimulate learning was the so-called programmed approach with projects having different timings compared to each other. The experimentation and replication of lessons between different individual projects can be classified as 'adaptive experimentation' (*ibid*). The sharing of lessons between projects has been facilitated by the PDR. The sequential phasing within individual projects, such as applied in the groyne-lowering project, can be considered a 'small-plot experiment' (*ibid*) that enabled replication within the same project.

None of the abovementioned authors refers to the importance of a culture of learning and the network to share lessons. In RftR we found the PDR deliberately creating a 'community' of involved professionals and politicians that documented lessons in guidelines and disseminated them through training and network events. The very existence of the PDR as 'lead-organisation' has stimulated community building. From this analysis it has been concluded that the RftR programme can be considered a lead organisation-governed network with high goal consensus (Provan and Kenis, 2007). The programme Directorate has been the lead organisation that, according to Graddy and Chen (2006), generally assumes most strategic and operational decisions. Folke et al. (2005) stress that the nodes (such as PDR) within the larger network of co-managers seem to be the main vehicle by which learning-by-doing occurs. Examining structural issues of a network (Provan et al, 2007) from the egocentric perspective of the PDR as lead organisation, we observed that the PDR has had a high and increasing: 'degree centrality' with many direct links to other organisations that participated in the programme such as municipalities and water boards; 'closeness centrality' having short paths to all other organisations in the network to spread knowledge; 'betweenness centrality' as a gatekeeper between other organisations such as between the national government and projects and amongst projects; 'multiplexity' with different types of links such as political, administrative, technical ties, research ties and shared personnel; and 'broker relationships' as the PDR could span gaps in the network. No examples have been found of PDR being connected to 'cliques' or clusters. Huntjens et al (2011) concluded that water management requires balancing between bottom-up governance (decentralisation) with top-down governance (decentralisation) for e.g.: facilitation, capacity building, cooperation across boundaries. The PDR provided a vehicle for this.

7.6 Discussion and conclusions

7.6.1 Learning outcomes are mutually enforcing

Using the categories of network learning, the learning outcomes of RftR have been classified unambiguously. Moreover, they enabled the analysis of interdependencies and feedback loops between the learning outcomes. The different types of learning outcomes as observed in RftR have been mutually enforcing. Lessons in terms of 'interpretations' have influenced changes in terms of 'structures' that, in turn, contributed to cognitive and behavioural learning outcomes in 'practices'. The new concepts of: river widening; integration of flood risk management and spatial planning; and programme management ('interpretations'), required collaborative arrangements ('structures') between national government and regional stakeholders; and enabled learning in participatory processes regionally ('practices'). Vice versa, learning in terms of practices has influenced structures and both have contributed to learning as 'interpretations'. When the monitoring results of the PDR indicated that many projects faced problems, the philosophy of 'controlled trust' was amended to a more collaborative approach and new organisational structures and processes were designed for facilitation of individual projects. Hence, we observe feedback loops between the learning outcomes that were confirmed during the network event to validate the findings. Knight and Pye (2004) already suggested their mutual dependence concluding that network outcomes follow a new alignment among network interpretations, structures and practices. This phenomenon is also implicitly highlighted through the feedback loops between single, double and triple-loop learning in social learning theories as described in Section 7.2. These feedback loops can provide an explanation why Armitage et al. (2008) concluded that it is difficult to distinguish different types of learning.

7.6.2 Network Learning helps to analyse both the adaptation of the physical system and of the societal system simultaneously.

The overlap and feedback loops between learning outcomes also indicate the relation between learning to adapt the physical system and the societal regime. Learning in RftR simultaneously supported the delivery of the programme and the contribution to a transition. River widening rather than dyke heightening and an adaptive co-management approach rather than a regulatory and technocratic approach required simultaneous and continuous learning and changes in interpretations, structures and practices during the programme. For example the concept of river widening guided the design and implementation of the individual measures that increase the river discharge capacity and subsequently lower water levels and decrease the flood risk. This new concept is also a demonstration of the transition to IFRM and will be considered as a measure for related and future policy and investment programmes, as shown by Van Herk et al. (2013c) based on a survey amongst 151 professionals in the Dutch water sector. The wider uptake of this concept is supported by various learning outcomes. Stakeholders developed new methods and tools such as the box of blocks and guidelines for planning, design and engineering that were necessary to deliver RftR, but will be used for other projects. More narrow collaboration between stakeholders and new collaborative arrangements were instrumental to deliver RftR, but also form the basis for collaborations in future IFRM programmes in the Netherlands such as the Delta Programme (Deltacommissie, 2011).

The Knight & Pye (2004) evaluation framework merits further validation, because it has shown potential to analyse simultaneously the contribution of RftR to: changes to the physical system (programme delivery); to the societal systems (outcomes); and, in addition, to adjustments of governance arrangements of the RftR programme itself (processes and structures: see next Section), all of which have been relevant to deliver IFRM in the case study. The results of this research indicate that for RftR the conceptual boundaries between transition management and adaptive co-management become opaque, because the learning and collaboration that took place in RftR is instrumental to both ACM and TM and the learning outcomes were generated simultaneously. Further case study comparison (preferably an international comparison) can help validate the wider applicability of the evaluation framework based on network learning. The framework might be able to support a structured analysis of various learning outcomes in a project or programme and their contribution to physical and societal system changes, as done for this case study. The research presented in this paper only used one case study and it was beyond its scope to include other case studies.

It is worth noting for the benefit of future research that learning outcomes in terms of 'structures' have only been studied at the level of an individual programme. Governance arrangements have been defined as the organisation of processes and structures within a programme. If the definition of governance arrangements is broadened to the organisation of the entire IFRM regime (i.e. the institutional and policy frameworks), then the evaluation framework can be used to explicitly study the impact of learning on changes to the 'structures' of the overarching societal system. Similarly, learning outcomes as structures can be studied for various functions of governance and phases of policy making such as formation, adoption and implementation (Teisman, 2000), rather than merely for programmes, which has been the focus of this research.

7.6.3 Programme and governance arrangements to stimulate learning and support adaptive co-management

Section 7.5 describes various generalised types of governance arrangements that have been proposed in a broad range of adaptive co-governance literature. Section 7.5 presents evidence that all of them have been applied to some extend in RftR. Case study comparison is necessary to investigate if and under which conditions programmes have the potential to adopt these types of governance arrangements: be managed at different relevant biophysical scale levels; be organised for polycentric governance; stimulate public participation; comprise different types of experimentation; and create a culture and network to stimulate learning and share lessons. More generically, further research is necessary to evaluate the potential of programmes to support adaptive co-management; and in turn to support management and adaptation of the physical and societal systems.

The research presented in this paper only used one case study and its results are not necessarily widely applicable. The governance arrangements applied in RftR were mentioned by interviewees, in the survey results and were confirmed by participants of the network event for validation as one of multiple reasons why the learning outcomes were generated. Interviewees mentioned: the integrated concept of river widening; the double objective of flood safety and spatial quality;

and the related design freedom and leadership for regional stakeholders to devise and implement measures, as other important catalysts of learning in RftR. In other words, if RftR had not adopted innovative integrated approaches from the outset, there might have been less and/or different learning outcomes, despite the organisation of the programme. Future research can investigate if the categories of learning outcomes as used in this paper can be converted into conditions to guide and evaluate the use of governance arrangements. A matrix can be created that confronts interpretations, structures, and practices with various ideal-type governance arrangements such as presented in Section 7.5. Also a list of key characteristics of programmes is to be developed to provide a basis for comparison, as it might be difficult to make a clear distinction between a programme and a project (see for a comparison: Maylor et al., 2006) or a policy (Rijke et al., in press). Following the definitions of programme and project as used in this paper, the RftR programme: comprised multiple projects; the projects were managed in a coordinated way; and the programme and projects had a defined beginning, end and other constraints. The key characteristics of the RftR programme in term of network learning were:

- Interpretations: a new integrated concept for the solution (river widening); new concepts on governance arrangements (regional leadership and programmed approach);
- Structures: multi-level network organization; with one centralised programme / network organization (Programme Directorate); flexible and adaptable structures and processes
- Practices: collaborative planning within design freedom

7.6.4 Adjust governance arrangements during a programme

The management of the RftR programme also used lessons learnt to adjust the governance arrangements of the Programme itself. The theory of network learning explicitly addresses this type of learning outcomes as 'structures'. It has been observed that learning outcomes in terms of interpretations and practices fed back into the programme and its governance arrangements. E.g. the interfaces between the Programme Directorate and the regional projects have been adjusted continuously in terms of meeting structures; monitoring; and facilitation. In particular the governance arrangements were modified responding to problems and challenges encountered during the implementation process. Such flexibility in adapting governance arrangements has been an important factor for successful implementation of RftR (see also: Rijke et al., in press). The Programme Director appositely explained the rationale behind his decisions on governance arrangements as: "structure should follow strategy". This research provides no evidence or recommendations on how to organise for flexibility other than a culture and leadership that stimulate learning; are receptive to lessons and are willing to consider changes.

The authors call for adaptive co-management and environmental management scholars to further theories by focusing more on programmes. Ideally evaluations are conducted during the implementation of programmes (*ex-durante*) and aim to actively contribute to the governance of these programmes. The adaptation of our physical systems will take place through projects and programmes, which in themselves also need learning for continuous adaptation of their governance arrangements to effectively deliver environmental change.

Chapter 8

Conclusions

Chapter 8 Conclusions

This research aimed to provide guidance on how to deliver integrated flood risk management. Following the research questions, the conclusions present: the definition of IFRM and a new framing of the inherent governance complexity (8.1); recommendations for governance arrangements to deliver IFRM, including a new process framework (8.2); and lessons on how projects can contribute to a transition to IFRM to enable a more widespread implementation (8.3). A reflection on two bodies of scientific literature is given in 8.4.

8.1 Framing the governance complexity of IFRM: contexts, processes, outputs, outcomes and feedback loops.

For this research, IFRM has been defined as an approach to develop and implement measures to reduce flood risk by collaboration between multiple disciplines; by a group of stakeholders with various interests and means; to combine objectives and funding from different policy domains; to consider a range of possible options at all spatial scale levels and for various time horizons (Chapters 3 and 4). This section will explain that, in addition to the development, implementation and operation of structural flood protection measures, IFRM considers additional measures and objectives; and requires the collaboration with additional stakeholders and disciplines. As a result, the governance complexity of IFRM is much greater than for traditional flood management alone.

The governance complexity has been framed to structure the guidance on how to deliver IFRM. Figure 8.1 graphically presents the conceptual framework that will be discussed in sequence from outputs, to processes, to context, to outcomes. The figure has been inspired by work of Pahl-Wostl et al. (2007). They focused on the role of learning to manage the physical system (natural and man-made) and change the societal system[11]. The figure has been enriched with a focus on projects[12] and has been validated based on evidence from all case studies in this research. The projects (multiple, overlaying boxes in the centre of Fig. 8.1 that are numbered 1, 2, and 'n' to indicate that multiple projects combined intervene in the physical system and influence the societal system) that have been studied: were

[11] A societal system is a part of society that can be attributed a functioning and functioning is the way a societal system meets a societal need. The functioning of societal systems can be described by its: structures, cultures and practices.

[12] A project is defined here as a temporary endeavour with a defined beginning and end (usually time-constrained, and often constrained by funding or deliverables, undertaken to meet unique goals and objectives, typically to bring about beneficial change or added value (based on Nokes, 2007).

conditioned by the social and physical systems (box in the upper part); created integrated outputs (boxes in the lower part) that changed the physical system; created outcomes that changed the societal system; and comprised planning processes based on collaboration and public participation; adaptive management; and learning.

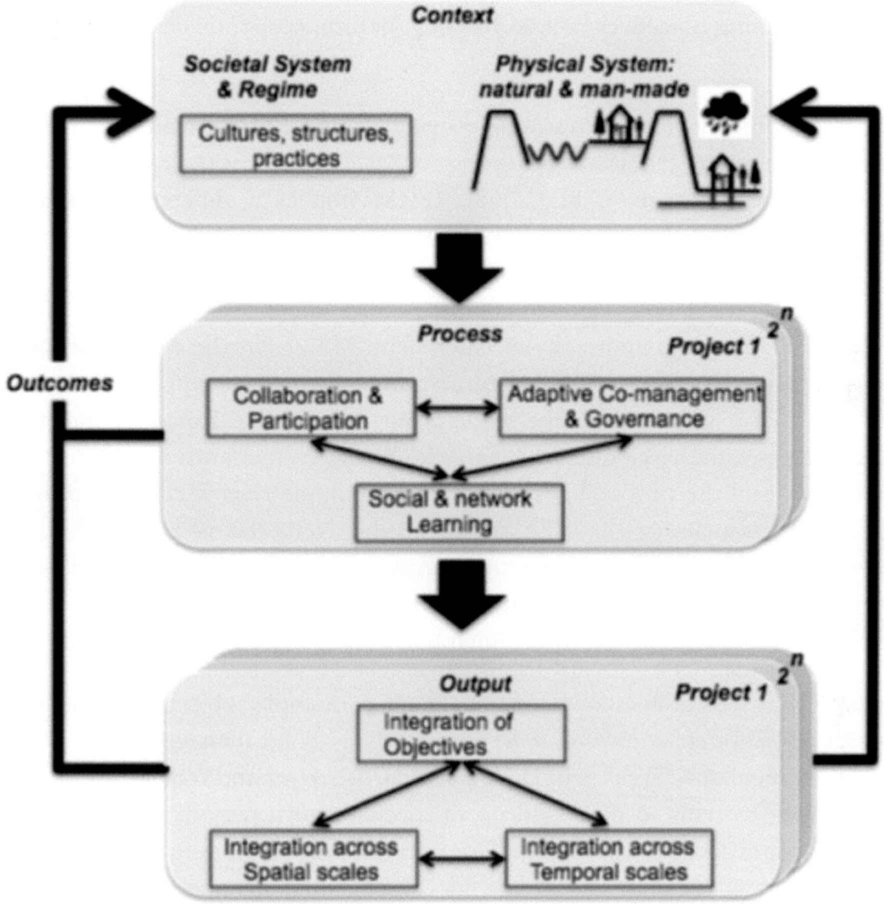

Figure 8.1. Conceptual framing of the complexity of IFRM governance

8.1.1 Outputs

The outputs of traditional flood management comprise structural flood protection measures such as a dyke or drainage system (as graphically represented in Table 8.1 by the figure in the first row) focussing on reducing the probability of occurrence of a flood event. The IFRM concepts that have been studied in this research better accommodate floodwaters using portfolios of measures (e.g. Evans et al, 2004), such as via river widening, urban development with flood proofing, increased storage capacity and/or green infrastructure, and/or embrace a range of soft- and non-structural measures. In IFRM these types of interventions are considered complementary or additional to structural flood protection measures. Consequently, the *objectives* of IFRM have broadened from reducing the probability of flooding, to reducing flood risk, which also includes reducing potential consequences of flooding necessitating a balance with other spatial and land use planning objectives such as housing, nature, economic development and transport.

The case studies from this research provide evidence that balancing multiple objectives inherently requires *integration across spatial scales*, as was already suggested by others (Zevenbergen et al., 2008). IFRM implies a different framing of problems and flood risk objectives at various scale levels and requires options at all spatial scale levels in the physical system to simultaneously: protect; prevent; and / or prepare e.g. individuals or communities, individual buildings, neighbourhoods, or an entire city or catchment. Moreover, the effectiveness of a given measure depends on other measures implemented at other spatial scales and levels that may be outside the scope of the project. E.g. Room for the River (RftR) increases the river discharge capacity in the Netherlands to reduce flood water levels and the probability of flooding of multiple river sections. However, the actual water levels that can occur in the Netherlands also depend on interventions taken upstream in Germany and Switzerland such as river widening and urbanisation in the flood plains that in turn influence how much water is actually discharged at the German-Dutch border into the Dutch river system.

Similarly this research shows that balancing multiple objectives inherently requires *integration across temporal scales.* Traditionally flood management has been based on a regulatory approach. Design standards are set and regulated for dykes and drainage systems to withstand up to specific return periods of rain or flow events or water levels in the rivers. Contrary, scenario planning that comprises inter alia the analysis of more extreme flood events than anticipated by the design standard has shown that the IFRM concepts as applied in the case studies, are more robust or provide flexibility to deal with climate change and related uncertainties regarding return periods, water levels and flows (Table 8.1 explains the Integration across temporal scales in the case studies). Combining or selecting measures from as wide a range of options as possible requires balancing short and long-term costs and benefits and anticipating (potential) future change and uncertainties (Adger et al., 2005; Zevenbergen et al., 2008; Haasnoot, 2013). Different options, such as flood protection systems, urban developments and emergency plans have different economic and technical lifespans, and their planning and operation and maintenance processes vary from each other.

The case studies from this research are listed in Table 8.1. It also presents the scope of traditional flood management to help understand the difference with IFRM that comprises a broader portfolio of measures. The images in Table 8.1 graphically represent the physical system in which IFRM can make interventions through measures. It comprises (from right to left): an urban area with: assets (represented by a house in the image), inhabitants (represented by an icon of a man) and nature (represented by a tree); a rain cloud that represents the hazard for pluvial flooding; an urban drainage system underneath the urban area to reduce the probability of pluvial flooding; a dyke that protects the urban area from fluvial and coastal flooding; the unembanked area or outer marches with assets, inhabitants and nature; the riverbed; and finally another dyke on the other side of the river. The scope of the proposed measures and of the IFRM concepts in the case studies is indicated with a circle. Table 8.1 also describes the integration in objectives (I^O) and across spatial (I^S) and temporal scales (I^T) as examined in Chapters 2, 3 and 4.

Table 8.1. The case studies from this research; their scopes and IFRM concepts; and the (planned) delivery of integrated outputs

Case Study	Scope & concept	(Planned) Delivery of integrated outputs: objectives (I^O), spatial (I^S) & temporal (I^T) scales
Traditional flood management (no case study)	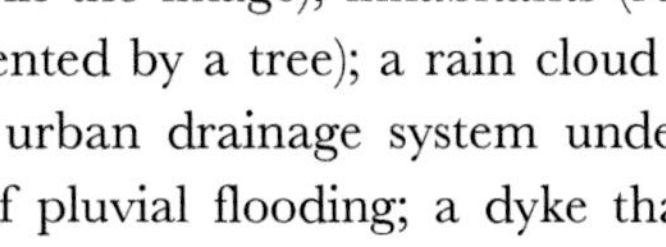	I^O: spatial integration of flood protection measures in the (surrounding) physical system necessitates consideration of local spatial objectives. I^S: Measures as part of a protection system for the entire river basis / coastal area / urban area; need local spatial integration. I^T: Performance of the flood protection up to a certain water level that has an estimated return period; possibly with headroom to protect lower probability water levels.

Stadswerven, flood proof urban development of a neighbourhood in the outer marches of Dordrecht

I^{O}: Residential and commercial development combined with flood proofing measures; to reduce individual flood risk and economic damage and to enhance the spatial quality of the neighbourhood.

I^{S}: Measures for individual dwellings and neighbourhood. Their performance depends on coastal protection and river discharge regime.

I^{T}: Considers the effectiveness of measures in reducing flood risk for various time frames under various climate scenarios. Gradual implementation over time provides flexibility to adjust plans.

Westflank, water storage capacity in regional urban development in Haarlemmermeer.

I^{O}: Water storage in high dense urban development in polder area to reduce the probability of flooding and potential damage and to enhance the spatial quality of the area.

I^{S}: Measures for individual dwellings; development of a regional spatial plan. Their performance depends on operation of the regional water system in adjacent areas.

I^{T}: Considers the effectiveness of measures in reducing flood risk for various time frames under various climate scenarios

Island of Dordrecht: multi-layered-safety (MLS) that combined flood protection, spatial planning and emergency management to increase the flood safety of the polder area protected by a ring dyke.

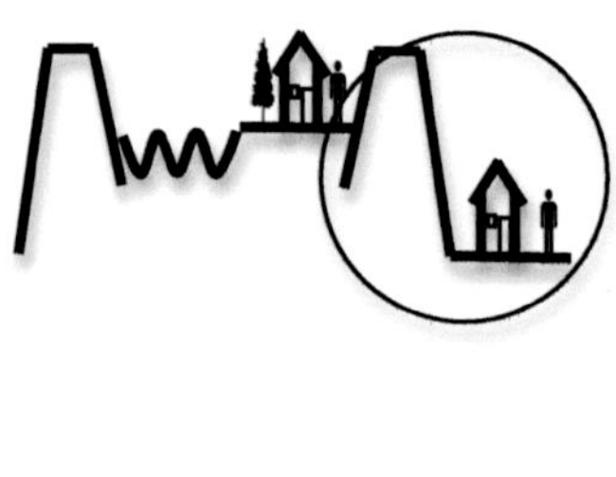

I^{O}: Increase flood safety by reducing individual, group and economic risk; combine several spatial developments.

I^{S}: Measures for dyke ring; city areas; critical infrastructure nodes and networks; and individual persons. Performance depends on

		coastal protection and river discharge; and on the occurrence of a flood event in adjacent areas to allow for evacuation from or to these areas. I^{T}: Considers different climate scenarios; various timings of investments in implementation paths (adaptation mainstreaming), flexibility to adjust strategy and measures.
Room for the River (RftR): delivered river widening measures to increase the river discharge capacity.		I^{O}: River discharge capacity; spatial development in floodplains and adjacent areas comprising: industrial; residential; nature; and recreation functions. I^{S}: Measures for the river system; river branches; and regional and local spatial developments. Perfomance dependent on international river basin management. I^{T}: Robust option for different climate scenarios; flexibility for future measures

The triplet of integration in objectives and across spatial and temporal scales has been used as an analytical classification for the case study research presented in this thesis. The classification is not to be misunderstood for an evaluation framework, although it can contribute as a tool to the development of a more comprehensive evaluation framework or collaborative evaluation process such as proposed by e.g. Licoln (1989) and Patton (2008; 2011).

The classification of integration does not imply a normative stance on the level of integration or on IFRM with respect to traditional flood management. An optimal strategy, if that can be defined and evaluated, can comprise one or multiple measures to reach a single or various objectives, for a single or various spatial scale levels and for a single or various time horizons. IFRM, as defined in this thesis, merely requires considering all options. Most of the literature used for this research (e.g. Huntjens et al., 2011; Pahl-Wostl et al., 2012) calls for IFRM and, often implicitly, takes a normative stance to IFRM following organisations such as the Global Water Partnership, although Jeffrey and Gearey (2006) argue that empirical evidence is missing that unambiguously demonstrates the benefits of these integrated approaches. The research presented in this thesis does not evaluate the benefits of IFRM and does not compare these to traditional flood management. In fact, the measures (e.g. dykes) and methods (e.g. flood probability analysis) of traditional flood management form part of IFRM that comprises a larger portfolio of measures and methods. This research has circumvented the normative debate by merely using case studies that have used an IFRM approach. The relevance of the research is illustrated by many new projects that, despite a lack of evidence of benefits (*ibid*) and clear guidance (Section 8.2), are now starting to aim for integrated processes and outputs (Butterworth et al., 2010).

8.1.2 Process

Flood risk management that integrates or balances between various objectives and across spatial and temporal scales, inherently comprises collaboration between multiple stakeholders and multiple disciplines (e.g. Potter et al., 2011). Governance arrangements have to stimulate the *participation and collaboration* of the various actors involved.

The tripartite: 'objectives-spatial scales-temporal scales' of IFRM and the multi-stakeholder process require *adaptive co-management and adaptive governance* as demonstrated in this research (Chapters 3 and 7). The various definitions from literature (Ruitenbeek and Cartier, 2001; Olsson et al., 2004; Folke et al., 2005; Dietz et al, 2003) all highlight the need for flexible governance and institutional structures for multi-stakeholder management and for learning to deal with the multi-objective reality and the dynamics of the physical systems. In this research, governance arrangements have been identified that provide flexibility for adaptive co-management and governance as a means to: enable integration of objectives across temporal scales (Chapters: 2; 3; and 4); deal with dynamics in the political and economic context; adjust for various planning phases; absorb lessons learnt (Chapters: 3 and 7); and adjust for and contribute to various transition patterns (Chapters 5 and 6). This research shows that the flexibility of governance arrangements has been an important factor in the delivery of IFRM in each of the case studies.

Hence, governance arrangements have to stimulate *learning*. A real case study or actual development project comprising an integrated concept (e.g. flood proof building or river widening); multiple objectives; participation of multiple stakeholders; and design freedom to explore alternative measures are the catalysts of learning (Chapter 7). They provide the context to work and learn together. Learning has been found to be essential to deliver the IFRM projects analysed and to support collaboration and participation. The case studies considered in this research have revealed that individuals and organisations have learnt from each other about their objectives, experience, disciplines, interests and means. Learning has been instrumental in exploring and assessing options and selecting and implementing measures to deliver the projects with various objectives at all spatial scale levels and for various time horizons. Moreover learning supports adaptive co-management and can generate outcomes that contribute to a transition to IFRM (Section 8.3). Various types of learning have been observed to have occurred or have been stimulated within various governance arrangements. Most authors of social learning (e.g. Pahl-Wostl, 2009; Armitage et al., 2008) adhere to a similar learning typology that distinguishes: single loop learning (refine actions); double loop learning (change values and policies); and triple loop learning (transform regimes) as developed by King and Jiggings (2002) and Keen et al.(2005). In this research also deutero learning (learn how to learn) (Tuinstra, 2008) and network learning in terms of practices (cognitive and behavioural), interpretations (philosophies or paradigms) and structures (adjust governance arrangements and patterns) (Knight and Pye, 2004) have been identified. Feedback loops have been observed between the types of learning and learning outcomes. Hence it is concluded here that it is difficult to distinguish between different types of learning, as has been previously hypothesised in literature (e.g. *ibid*; Armitage et al., 2008).

The three process features are mutually reinforcing (Chapters 2; 7). Collaboration and participation leads to learning, and learning supports collaboration. Adaptive co-management comprises collaboration by definition and requires learning and feedback loops to deal with the dynamics of: the physical system; the political and economic context; planning phases and transition patterns. A framework has been developed based on the research presented in this thesis to analyse and structure collaborative planning for IFRM that adopts the process features: collaboration & participation; adaptive co-management; and learning (Section 8.2.2).

8.1.3 Context and outcomes

IFRM projects cannot be seen in isolation from their wider societal and physical contexts. Policy and investment projects, such as studied in this research, are interventions that change the physical conditions and together adapt the physical system. In addition, the projects are to be delivered within an overarching context of an incumbent societal system comprising institutions and policies that are not (yet) designed for IFRM. This research provides guidance on how projects can contribute to a transition that changes structures, cultures and practices of the societal system to enable a more widespread implementation of IFRM (Section 8.3). The projects have contributed to the transition by generating outcomes beyond the scope of the project and that can be sustained beyond the duration of the project. Governance arrangements aim to stimulate the delivery of projects within their contexts, and ideally also stimulate the generation of outcomes to change the regime.

8.2 Governance arrangements to deliver IFRM projects

Literature provides limited guidance on the use of governance arrangements to organise a collaborative planning processes to deliver IFRM (Chapters 2 and 3). Few examples of such processes have been identified, in practice, documented and evaluated (Pahl-Wostl et al., 2007) as the implementation of truly *integrated* FRM plans is still in its infancy (Huntjens et al., 2011). In recent years, in parallel with this research, much other research has been conducted that also provides insights on how to deliver IFRM, but each research had a different emphasis. Bos and Brown (2012) and Farrelly and Brown (2011) use elaborate case study research that focus on governance arrangements for experimentation to stimulate social learning and to change the societal system, rather than focus on developing and implementing IFRM plans. White (2008), Zevenbergen et al. (2010) and Van de Ven et al. (2011) present a range of possible IFRM measures and activities, but do not provide ideas on how to organise these activities into a coherent process or lack empirical evidence (Chapter 3).

This research focuses on real-life case studies in which IFRM plans have been developed or implemented through projects that comprised interventions to adapt the physical system. From this research attributes for IFRM projects have been deduced to foster integration and implement the projects (Section 8.2.1). Moreover a framework has been developed that has been useful, both descriptively as an analytical tool and prescriptively to structure collaborative work and to deliver IFRM outputs in the case studies (Section 8.2.2). The framework adopts the process features: collaboration & participation; adaptive co-management; and learning (as presented in Section 8.1). Multiple authors (e.g. Armitage et al., 2008) provide frameworks to stimulate and evaluate on each separate feature from various scientific domains, but not in a practical integrated manner, and mostly to manage and change the physical or the social system rather than on the level of individual projects. The results of the case studies in this research indicate that for projects, the process features are mutually reinforcing and can be pursued simultaneously by governance strategies.

8.2.1 Attributes to foster and implement IFRM

This research has shown that IFRM projects require fit-for-purpose (see also: Rijke et al. 2012a) and fit-for-context governance arrangements. Some common lessons have emerged from this notion that have resulted in the development of a framework. From the research two types of attributes for IFRM projects have been deduced: those that directly foster IFRM processes and outputs and those that enable the implementation of IFRM projects (Chapter 4). The attributes to foster IFRM processes and outputs that have been deduced are:

- Start with an integrative vision and concept;
- Set multiple objectives;
- Provide (delimited) design freedom;
- Organise for mutually enriching planning activities; and
- Organise for collaborative planning & public participation.

Attributes to support the implementation of IFRM projects that have been deduced are:

- Disasters and failures of sectoral approaches are to be exploited to adopt an integrated concept and approach and set multiple objectives.
- Integrated outputs are to be anchored through committed sponsors who assign resources through formal agreements, or otherwise to be stimulated by committed advocates.
- The collaborative planning process is to establish legitimacy, build trust and embed accountability systems to safeguard support for the integrated process and outputs throughout the planning process.
- The collaboration structure is to be fit-for-purpose and adaptable: balance top-down and bottom-up governance; and formal and informal relations.
- The planning process needs to stimulate learning and feedback loops to absorb threats and seize opportunities from a dynamic context.

8.2.2 A new framework to analyse and structure collaborative planning in IFRM

A framework has been developed to analyse and structure collaborative planning. It has been developed based on the Stadswerven and Westflank case studies and validated and has been enriched based on the MLS and RftR case studies. The framework organises project activities (that can comprise different work tasks) along three lines: system analysis; collaborative planning, design and engineering; and governance. These activities generate knowledge (establish facts; create images and set ambitions) and support decision making (address problems; develop solutions and involve participants and influence politics). The activities can be analysed or (re-)designed successively for different planning phases. Figure 8.2 illustrates the framework with the 3 activities. The 3 activities (which actually comprise matrices) are presented in a loop and are interconnected by arrows as the activities are mutually supportive and run in parallel. Each activity is represented as a matrix that shows how the activity delivers in regard to generating knowledge and supporting decision-making in 5 cells. The interactions can be deduced from Fig. 8.2 by overlaying the matrices of each activity, with overlaps occurring for four cells in each activity.

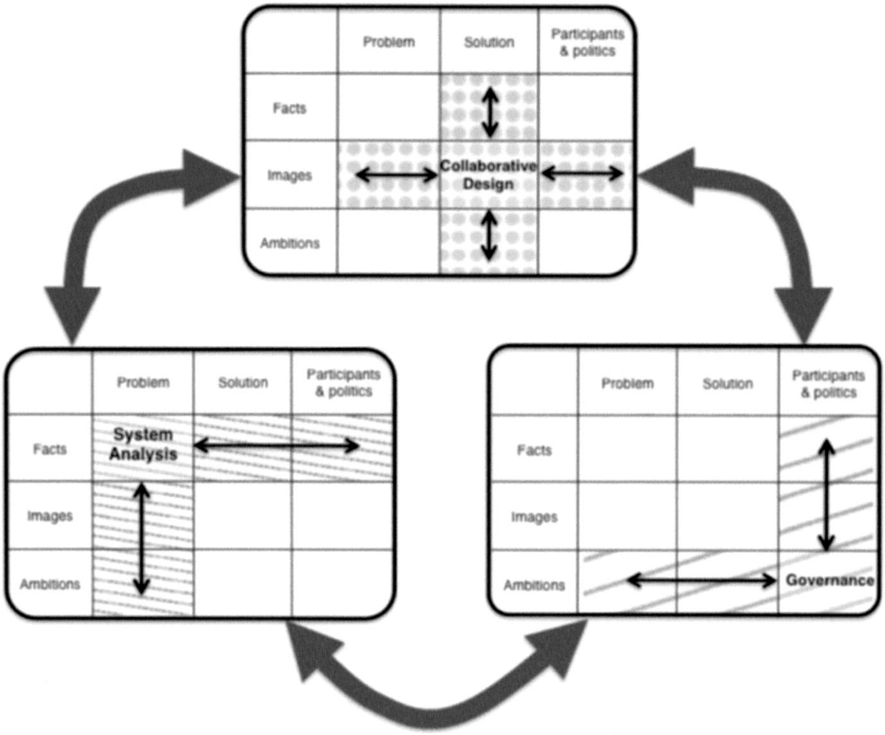

Figure 8.2. 3 interactive IFRM activities to generate knowledge (facts, images, ambitions) and support decision making (problem, solution, participants & politics), from Van Herk et al. (2011a)

This research has provided evidence on how the activities can contribute to integrated outputs, i.e. to the measures and strategies (to be) implemented and their integration in objectives, across spatial and temporal scales (Chapters 3 and 4). The planning process needs to be granted design freedom. Design freedom provides the scope to devise integrated outputs within an integrated concept, such as river widening or flood-proof urban development. The activities are to be mutually enriching. They undergo a process of divergence (knowledge development in one activity) and convergence (knowledge exchange between activities) in each planning phase. The mutual enrichments of proposing (designing), analysing and selecting (governance) contributes to the integrated nature of the outputs (Chapter 3).

The framework has been enriched based on the MLS case study (Chapter 3) by including phasing during the process of developing an IFRM plan (but not its implementation) and by providing examples of sub-activities during various planning phases, within the 3 categories of system analysis; planning, design & engineering; and governance. The observed phasing relates to the process of divergence and convergence that has been explained by the interaction between activities. In Figure 8.3 the process is illustrated; with circles representing the three activity types of the framework (system analysis; planning, design and engineering; and governance); and 9 sub-activities that have been conducted in 3 different phases, either within, or on the interface of the 3 activity types. In the first phase (phase 1) of 'divergence': the system performance is analysed, e.g. in terms of the probability and potential consequences of flood events; stakeholders are brought together and set their objectives; and different strategies, options or measures to intervene in the flood risk management system[13] are explored. Bringing the outcomes of these series of activities together on the interfaces between activity types leads to a new phase (phase 2) of: problem (re)framing or joint goal-setting based on the discussion of the system performance and different objectives; to discuss strategies and options based on the combined means and objectives of all stakeholders; to assess the performance of identified strategies and options, also using the objectives discussed. Ultimately (phase 3) the options are to be combined into an IFRM plan comprising a portfolio of measures and an investment and implementation plan; that is subject to formal decision-making processes between the stakeholders and within the democratically representative bodies of governmental organisations involved. Also the performance of the IFRM plan is to be analysed, which inherently is a feedback loop to the continuous monitoring of system's performance (phase 1) that will change with the implementation of the IFRM plan and due to exogenous factors (e.g. Milly et al., 2008). The set of sub-activities in the first phase can be conducted in parallel as they are independent. The sub-activities in phase 2 and 3 have feedback loops, just as there can be feedback loops and iterations between the phases. Hence, the enriched framework as presented in Figure 8.3 can provide guidance for the

[13] The whole of the physical systems, actors and rules required to manage flood risk. (Gersonius, 2012).

design and coordination of the collaborative planning process to develop an IFRM plan, but is not a blueprint process design. Neither does the framework guide further planning phases towards implementation. Based on the case study RftR, a first attempt has been made in this research to further detail the framework for planning phases after the development of an IFRM plan (Van Herk et al., 2012a, not included in this thesis). The activities during the planning phases design, construction and operation could be classified unambiguously following the three activity types. However the case study results from RftR require theoretical conceptualisation; validation and case study comparison to further enrich the framework for all planning phases.

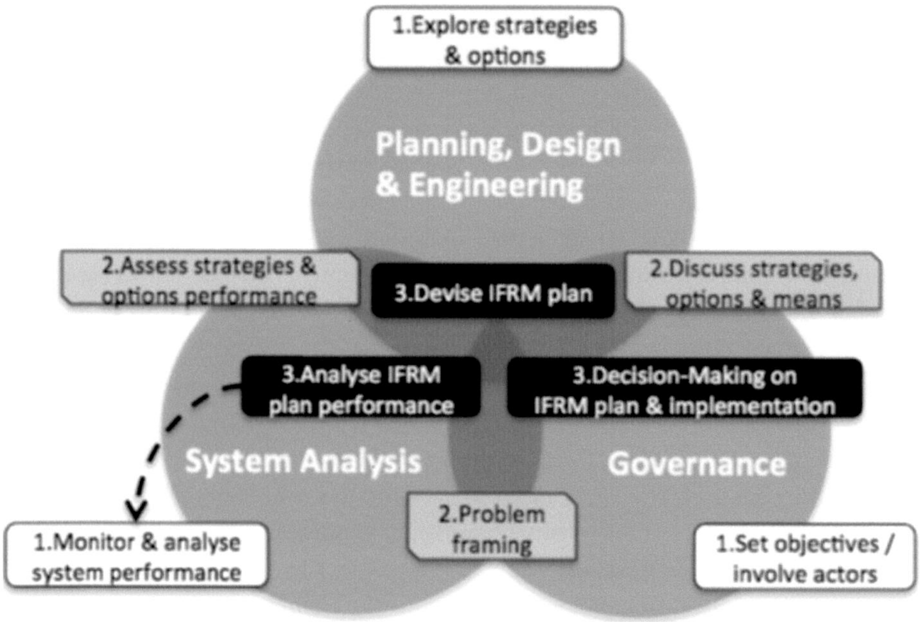

Figure 8.3. Adapted framework for collaborative planning processes to deliver IFRM plans focusing on interfaces between activities, from Van Herk et al. (2013a)

Further research is recommended into the management of governance activities and the definition of an evaluation framework for the governance activities. For the case studies Stadswerven, Westflank and the Island of Dordrecht, an IFRM plan has been developed based on the process framework, but implementation is uncertain due to a lack of legitimacy and commitment of the collaborating stakeholders (Chapters 2 and 3). The governance activities are not only to engage a broad range of stakeholders to define ambitions and select strategies, but also to explore their legal and other mandates and willingness to commit or combine resources. The governance activity can comprise different aspects (as adapted from Voß and Bornemann, 2011): 'policy' (discussing problems and solutions); 'polity' (rules and structure); and 'politics' (interaction and process). Using these definitions, the case studies focused mostly on policy and much less on polity and politics.

Lessons have emerged and frameworks have been developed and validated for analysing and structuring collaborative planning in IFRM, but a more detailed blueprint for governance arrangements for this cannot be devised from the data. Several reasons have been found for this. All case studies from this research are situated in the Netherlands and the transferability of lessons depends on the: geographical; historic (flood history); cultural and socio-economic; and institutional contexts (Klijn et al., 2008; Zevenbergen et al., 2013a). The case studies that have been analysed started with different problem framings; with different types of funders and primary objectives; and with different integrated concepts (See Table 8.1). The framework needs to be validated across different IFRM projects in different countries, to further its theoretical development and applicability. In practice, a customised application of the framework is hypothesised to be necessary for each project.

8.3 Governance arrangements for projects to deliver in, and contribute to a changing regime

Many scholars call for a transition to societal systems that actively manage flood risk to reduce flood impacts and accommodate floods: 'living with water', in addition to a mere focus on flood protection: 'fighting against water' (e.g. White, 2010; Newman et al., 2011; Dawson et al. 2011; Zevenbergen et al., 2013b). The relation between the projects and the changing societal system is two-way. The projects researched in this thesis have taken place during; have been a part of; and are a demonstration of this transition (Chapters 2; 3; 5; 6). Conversely, this research has also shown that the actual implementation of the projects supported the adaptation of the regime to the new paradigm (Chapters 5 and 6). This research builds upon and enriches existing frameworks from transition literature that to date have not been applied to analyse the impact of individual projects, nor to water management (Chapter 6). This has provided in-depth evidence of how the transition to IFRM has occurred in practice. Lessons have been deduced on governance arrangements that are to stimulate the delivery of projects within their contexts, and also stimulate the generation of outcomes to change the regime.

8.3.1 Deliver an IFRM project in an *unsympathetic* and changing regime

The case studies included in this research demonstrate the transition the societal system has been going through from a flood protection regime that dominated the functioning of the system, to an IFRM regime. All projects studied for this research started with new, integrated concepts that embody the new paradigm of living with water: river widening (RftR); flood proofing (Stadswerven); building with water storage (Westflank); and multi-layered-safety (Island of Dordrecht). The case studies exemplify three ideal-type transition patterns developed by De Haan and Rotmans (2011) that together describe how the societal system and the regime changed: top-down reconstellation; bottom-up empowerment; or adaptive and internally induced. The projects Stadswerven, Westflank and Island of Dordrecht exemplify smaller scale initiatives that can become viable alternatives to mainstream daily practice (Chapters 2; 3; and 5). The RftR programme can be seen as part of the empowerment of the small constellation[14] that started with many small initiatives that gained power and were scaled up and incorporated in national policies of the incumbent regime. Alternatively, the launch of the RftR programme can be considered a reconstellation or adaptation pattern after a 'top-down' change of policy by the national Government and Parliament (Chapter 6).

The projects that have been studied were to be delivered within an overarching context of incumbent societal systems comprising institutions and policies that were not (yet) designed for IFRM. The different collaborative frameworks that have been used (collaborative research; Learning and Action Alliances (LAAs); or an embedded institution such as RftR's programme directorate) were designed and selected to provide flexibility to deal with this challenge. E.g. the Stadswerven project was the very first project that considered flood proof building in outer marches and policy and regulation that was in force at the time of the project did not permit this. The project did not have an official experiment status that would allow it to circumvent the legal limitations. Collaboration between various stakeholders was organised in a research project that provided the freedom to stakeholders to explore options for flood proofing unconstrained by legal boundaries or political positions. The options could later be demonstrated and implemented in practice. To deliver RftR the national government empowered regional authorities with the planning and implementation of the projects, as opposed to mandating this to Rijkswaterstaat, the executive arm of the Ministry of

[14] A societal subsystem. "The constellation that dominates the functioning of the system will be denoted as the regime" (De Haan and Rotmans, 2011, pp 93).

Public Works, which was the prevailing practice to implement large infrastructure programmes in the Netherlands (Hertogh et al., 2008). RftR used a new organisational framework to organise, monitor and facilitate collaboration between national and regional authorities and between individual projects to stimulate progress of the programme and address governance challenges that might become apparent. Moreover the framework was flexible and has been adapted continuously based on lessons learnt.

8.3.2 Project contribution to regime change

The projects have contributed to the transition by generating outcomes beyond the scope of the project and that are sustained after the duration of the project. Various types of outcomes have been found. New options (such as river widening and flood proofing) and related practices and methods for planning, design and analysis of these options have been developed and demonstrated that can be replicated elsewhere. Policy and regulation has been adjusted to allow for the implementation of new options in the case studies and for future projects. For example: to allow for the development of spatial functions in outer marches, areas unprotected by dykes in the riverbed (Chapters 2 and 5); to enable the movement of soil that has been excavated for river widening and subsequently the use of this soil for various new functions; creating precedents for dyke design and engineering (Chapter 6). New governance arrangements such as the LAAs for De Stadswerven or a Programme Directorate for RftR are being used to organise future projects and programmes (Dudley et al., 2013; Chapter 6). The projects contributed to capacity building and creation of networks amongst individuals and organisations that have been involved in the case studies (Chapters 5 and 6). They have indicated that this will support them to implement future IFRM policies and projects.

The outcomes explain the conditions for transitional change that De Haan and Rotmans (2011) have classified in: pressure from alternatives; structural and cultural tension from the environment; stress from internal inconsistencies. Vice versa, the transition conditions explain the outcomes that have been generated (Chapter 6). All case studies started with an integrated concept that comprised alternative or complementary options to traditional flood protection options (Table 8.1). These alternatives put pressure on the incumbent regime. Cultural tension emerged because politicians and professionals were not used to working with an IFRM approach, which required capacity building, new practices and methods. Current policy and regulation that was based on a flood protection approach created structural tension and hampered the implementation of IFRM options for which adjustments of policy and regulation were necessary. The institutional context, collaborative structures and project management approaches that prevailed were based on a sectoral approach and unfit for IFRM. New governance arrangements have been used to overcome this stress.

Learning has been important to generate the project outcomes and overcome the conditions for transitional change in the case studies (Chapters 5 and 6). The governance arrangements that were adopted deliberately stimulated learning. The Learning and Action Alliances (LAAs) for the cases Stadswerven, Westflank and Island of Dordrecht had the freedom to address new and politically sensitive problems and explore innovative solutions, unconstrained by incompatible regulation or formal political positions and with participation of a variety of stakeholders and disciplines. LAAs had as an explicit objective to scale-up their innovative approaches and contribute to a transition. RftR used a programmed approach and national-regional agreements with a Programme Directorate as a node to facilitate collaboration and learning. Moreover, RftR created a culture of learning and a network to disseminate lessons that implicitly supported the generation of outcomes.

8.4 Merge theories on adaptive co-management and transition management for IFRM projects

This research has drawn upon two bodies of literature (inter alia): adaptive co-management (ACM) and transition management (TM). ACM studies the physical system that is changed by the implementation of IFRM projects. TM studies changes of societal systems[15] that are required for and furthered by the implementation of IFRM projects. Both bodies of literature have similar theoretical foundations (Van der Brugge and Van Raak, 2007) and multiple authors (*ibid*; Smith and Stirling, 2010; Voß and Bornemann, 2011; Armitage, 2008; Pahl-Wostl et al. 2010) have made attempts to combine and mutually enrich their theoretical concepts. Here the definitions of the physical system and societal system and the description of the scope of ACM and TM have been simplified. The two bodies of literature also use various other, overlapping terminology, such as social-ecological system (the co-evolutionary units of social and ecological systems (Folke et al, 2005))and socio-technical system (all the physical systems, actors and rules required in order to perform a particular

[15] A societal system is a part of society that can be attributed a functioning and functioning is the way a societal system meets a societal need. The functioning of societal systems can be described by its: structures, cultures and practices.

function (Geels, 2005)). These terminologies combine 'ecological' or 'technical' with 'social' to indicate the powerful reciprocal feedback loops between the systems. Hence, the attempts of abovementioned authors to develop theories that comprise both the adaptation of the physical and societal system as defined in this research. However, neither body of literature provides guidance for ACM or TM on a project level, but rather conceptualised theories at the system level, whilst this research shows that ACM and TM can be merged and applied simultaneously for IFRM projects (Chapter 7).

ACM and TM are both learning-oriented management frameworks that stress the importance of learning through collaboration between the various stakeholders that are involved in or affected by the management and change of the physical and societal systems. Various authors have developed conceptual frameworks to classify the goals, outcomes and approaches to various types of learning by individuals, organisations and networks (e.g. Armitage et al., 2008; Steyaert and Jiggins, 2007; Pahl-Wostl, 2009; Folke et al, 2005; Huitema et al., 2009). They focus either on learning to adapt the physical system to uncertainty and changes (ACM), or on learning to adapt the societal systems (TM). Social learning is presented as a means to adapt and deliver change and the authors adhere to the learning typology that distinguishes: single loop learning; double loop learning; and triple loop learning (Section 8.1.2). It is argued in this research that this typology is used to describe learning at the level of systems or sectors (e.g. water management), but that none of the literature mentioned above explicitly includes learning in individual projects or programmes. It is complicated to evaluate learning in projects based on the loop-typology, because there are many different definitions and meanings that complicate the task of defining common indicators to measure social learning as either a process or an outcome (Muro and Jeffrey, 2008; Armitage, 2008).

The research described in this thesis has taken 'projects' as the level of analysis that has been overlooked to date in ACM and TM despite its significance to explain system changes. The pro-active adaptation of the physical system (ACM) takes place through interventions that are part of different policy and investment projects, such as the case studies in this research, that together form adaptation pathways (Gersonius, 2012). Also, the case study results showed how projects support transition patterns that change societal systems and regimes (TM). The focus on the 'project' as taken here, can further the development and merger of ACM and TM based on the empirical evidence from this research. The results of this research indicate that the conceptual boundaries between ACM and TM become opaque for projects, because the learning and collaboration that takes place in projects is instrumental to both ACM and TM and the learning outcomes are generated simultaneously (Chapter 7). This research has drawn upon a learning theory from a different domain, namely health services, to evaluate learning in projects.

An evaluation framework based on network learning (Knight and Pye, 2004) has been applied and enriched in this research that can support a structured analysis of various learning outcomes of projects and their contribution to changes to both the physical and societal system. Knight and Pye (2004) use network learning, an alternative theory to the loop typology, to classify learning outcomes into 3 categories: interpretations (philosophies or paradigms); structures (adjust governance arrangements and patterns); and practices (cognitive and behavioural). Using these categories, the learning outcomes of RftR have been classified unambiguously. Moreover, this has enabled the analysis of interdependencies and feedback loops between the learning outcomes (Chapter 7).

The Knight & Pye (2004) evaluation framework merits further validation, because it has shown potential to analyse simultaneously the contribution of projects to: changes to the physical system (outputs); to the societal systems (outcomes); and, in addition, to adjustments of governance arrangements of projects themselves (processes and structures), all of which have been relevant to deliver IFRM in the case studies. Examples of changing interpretations comprise the increased awareness of and support for integrated concepts such as river widening (RftR) or flood proofing (Stadswerven), building with water storage (Westflank) and multi-layered-safety (Island of Dordrecht). These concepts have been the basis for the measures that have been developed and are being implemented to change the physical system. They have demonstrated transition patterns that either empower an alternative approach or require adaptation of the societal system. These inherently required and stimulated collaborative and flexible governance arrangements. The LAAs and the flexible programme organisation of RftR are examples of structures that have been (re)designed based on learning (Chapters 2 and 7). They have been designed and adjusted to enable the participation and collaboration of multiple stakeholders to deliver IFRM. They contributed to a transition by generating outcomes such as new networks of individuals and organisations and their capacity to collaborate effectively within these new collaborative governance arrangements. The structures provided flexibility to adjust governance arrangements for various planning phases; to enable integration across temporal scales; and deal with dynamics in the political and economic context. Learning in terms of practices has contributed to the delivery of IFRM by working together on collaborative design, analysis and decision-making. It has generated outcomes beyond the scope and duration of the projects via capacity building of individuals and organisations and by the creation of guidelines. Governance arrangements have been adjusted based on lessons learnt during the processes, such as engaging new stakeholders like decision makers or project managers in the LAAs (Chapters 2 and 3) or a shift from monitoring of regional authorities to facilitation of them by national government in RftR (Chapter 7).

Chapter 9 References

Adger W.N., Arnell, N.W., Tompkins, E.L. 2005. Successful adaptation to climate change across scales. Global Environmental Change, 15, 77-86.

Agranoff, R. & McGuire, M. 2001. Big questions in public network management research. Journal of public administration research and theory, 11(3), 295-326.

Agrawal, A. 2003. Sustainable Governance of Common-Pool Resources: Context, Methods, and Politics. Annual Review of Anthropology 32, 243-262.

Albers, J. 2012. Waarborging van ruimtelijke kwaliteit in Ruimte voor de Rivierprojecten: in de overgang van planstudie- naar uitvoeringsfase, Master thesis, Erasmus University Rotterdam.

Armitage, D., Marschke, M., Plummer, R. 2008. Adaptive co-management and the paradox of learning. Global Environmental Change 18, 86–98.

ASCE American Society of Civil Engineers 2011. Policy Statement 511 - National Levee Safety Program

Ashley, R.M. & Cashman, A. 2006. The Impacts of Change on the Long-term Future Demand for Water Sector Infrastructure. Infrastructure to 2030: telecom, land transport, water and electricity , OECD, ISBN 92-64-02398-4

Ashley, R.M., Blanskby, J., Newman, R., Gersonius, B., Poole, A., Lindley, G., & Nowell, R. 2012. Learning and Action Alliances to build capacity for flood resilience. Journal of Flood Risk Management, 5(1), 14-22.

Barnett J. & O'Neill S. 2010. Maladaptation. Editorial for Global Environmental Change 20, 211-213. doi:10.1016/j.gloenvcha.2009.11.004. Elsevier.

Baskerville, R.L. and Wood-Harper, A.T. 1996. A critical perspective on action research as a method for information systems research. Journal of Information Technology 11, 235-46.

Batchelor C. & Butterworth J. 2008. Learning Alliance Briefing Note 9: Visioning (draft). http://www.switchurbanwater.eu/la_guidance.php [accessed November 2008]

Belt, M., Bowen, T., Slee, K., & Forgie, V. 2013. Flood Protection: Highlighting an Investment Trap Between Built and Natural Capital. JAWRA Journal of the American Water Resources Association 49, 681-692.

Berkes, F. 2009. Evolution of co-management: Role of knowledge generation, bridging organizations and social learning. Journal of Environmental Management 90, 1692–1702.

Biswas, A.K. 2004. Integrated water resources management: a reassessment: a water forum contribution. Water international, 29(2), 248-256.

Bläckler, F. 1995. Knowledge, knowledge work and organizations: an overview and interpretation. Organizational Studies 16, 1021-1046.

Blackmore, C., Ison, R., & Jiggins J. 2007. Social Learning: an alternative policy instrument for managing in the context of Europe's water. Environmental Science & Policy 10, 493 -586.

Blom, E.T., Zevenbergen, C., Gersonius B. & Van Son E. 2012. Smart shelter strategies; cost-effective flood preparedness. Floodrisk 2012, Rotterdam, Netherlands.

BMW, 2008. Bouwen met Water, www.bouwenmetwater.nl [accessed 18/02/08]

Boelens, L. 2006. Beyond the Plan; Towards a New Kind of Planning. disP ,167, 4, 25-40.

Bos, J.J., & Brown, R.R. 2012. Governance experimentation and factors of success in socio-technical transitions in the urban water sector. Technological Forecasting and Social Change, 79(7), 1340-1353.

Bosher L. & Coaffee J. 2008. Editorial: International Perspectives on urban resilience. Proc. of the Institution of Civil Engineers Urban Design and Planning 161, 145-146.

Bouwen, B., & Taillieu, T. 2004. Multiparty collaboration as social learning for interdependence: developing relational knowing for sustainable natural resource management. Journal of Community and Applied Social Psychology 14, 137-153.

Bouwer, L.M., Crompton, R.P., Faust, E., Höppe P., Pielke, R.A. 2007 Confronting disaster losses. Science 318, 753.

Bozeman, B. 2000. Technology transfer and public policy: a review of research and theory. Research Policy 29, 627-655.

Breit, H., Engels, A., Moss T., Troja, M. 2003. How institutions change: perspectives on social learning in global and local environmental contexts. Leske + Budrich, Poladen, Germany.

Brower, R.S., Abolafia, M.Y., & Carr, J.B. 2000. On improving qualitative methods in public administration research. Administration & Society 32, 363–97.

Brown, R.R., Keath, N., Wong, T. 2008. Transitioning to Water Sensitive Cities: Historical, Current and Future Transition States. Proc. 11th Int. Conf. Urban Drainage. Edinburgh.

Brown, R., Ashley, R., & Farrelly, M. 2011. Political and professional agency entrapment: an agenda for urban water research. Water Resources Management 25, 4037-4050.

Bryson, J.M., Crosby, B.C., & Stone, M.M. 2006. The Design and Implementation of Cross-Sector Collaborations: Propositions from the Literature. Public administration review 66, 44-55.

Butterworth, J., Warner, J., Moriarty, P., Smits, S., & Batchelor, C. 2010. Finding practical approaches to integrated water resources management. Water Alternatives, 3, 68-81.

Carter, N., Kreutswizer, R.D. & de Loë, R.B. 2005. Closing the circle: linking land use planning and water management at the local level. Land Use Policy 22, 115–127.

Chalmers, A.F. 1982, What is this Thing called Science?, University of Queensland Press, Brisbane.

Checkland, P. 1991, From framework through experience to learning: the essential nature of action research, in Nissen, H.E. et al. (Eds), *Information*

Systems Research: Contemporary Approaches and Emergent Traditions, Elsevier, Amsterdam.

Checkland, P.B. & Holwell S.E. 1998. Action research: its nature and validity. Systemic Practice and Action Research 11, 9–21.

Cleaver, F. 2002. Reinventing institutions: bricolage and the social Embeddedness of Natural Resource Management. The European Journal of Development Research 14 , 11-30.

Commissie Duivesteijn, 2004. Grote projecten uitvergroot, Een infrastructuur voor besluitvorming

Commissie Elverding, 2008. Sneller en beter - Advies Commissie Versnelling Besluitvorming Infrastructurele Projecten.

Cook, S.D.N. & Brown, J.S. 1999. Bridging epistemologies: the generative dance between organizational knowledge and organisational knowing. Organization Science 10, 381-400.

Cook, W., Casagrande, D., Hope, D., Groffman, P. & Collins, S., 2004. Learning to roll with the punches: adaptive experimentation in humandominated systems. Frontiers in Ecology and Environment 2, 467–474.

Cornwall, A., & Jewkes, R. 1995. What is participatory research? Social Science & Medicine 41, 1667-1676.

Cuppen, M.E. 2007. The problem of legitimacy in urban flood management redesign processes. In International Conference on Adaptive and Integrated Water Management. Basel. Switzerland (pp. 12-15).

CW21. 2000. Anders Omgaan met Water. Waterbeleid voor de 21e eeuw. (Dealing Differently with Water, Water management for the 21st century). Committee Tielrooy, Ministry of Transport, Public Works and Water management/Union of Waterboards. The Hague, The Netherlands.

Daniell, K.A., White I., Ferrand, N., Ribaravo, I.S., Coad, P., Rougier, J.E., Hare, M., Jones, N.A., Popova, A., Rolling, D., Perez, P. & Burn, S. 2010. Co-engineering Participatory Water Management Processes: Theory and insights from Australian and Bulgarian Interventions. Ecology and Society 15, 11.

Dawson, R.J., Ball, T., Werritty, J., Werritty, A., Hall, J.W., & Roche, N. 2011. Assessing the effectiveness of non-structural flood management measures in the Thames Estuary under conditions of socio-economic and environmental change. Global Environmental Change 21, 628-646.

DCLG 2009. Planning Policy Statement 25: Development and Flood Risk – practice guide. Department of Communities and Local Government, London ISBN: 978-1-85112-936-2.

DCLG 2012. National Planning Policy Framework. Department for Communities and Local Government, England. March. ISBN: 978-1-4098-3413-7 www.communities.gov.uk.

DCLG 2012b. Technical Guidance to the National Planning Policy Framework. ISBN: 978-1-4098-3410-6. Crown copyright

De Bruijn, J.A. & ten Heuvelhof, E.F. 1999. Scientific expertise in complex decision-making processes. Science and public policy 26, 179-184.

De Bruijn, K.M. 2005. Resilience and flood risk management, a systems approach applied to lowland rivers. Dissertation, Delft University of Technology

De Haan, J. & Rotmans, J. 2011. Patterns in transitions: Understanding complex chains of change. Technical forecasting & social change 78, 90-102.

De Kraker, J., Cörvers, R., Valkering, P., Hermans, M., Ruelle, C. 2010. Regional learning networks for sustainable development. Presented at ERSCP-EMSU 2010 Conference: Knowledge Collaboration & Learning for Sustainable Innovation, Delft (The Netherlands)

Deltacommissaris, 2011. Deltaprogramma 2012 - Werk aan de delta. Ministerie van Infrastructuur en Milieu, Report, Deltacommissie, The Hague

Deltacommissie, 2008. Working together with water; A living land builds for its future; Findings of the Deltacommissie 2008, Technical Report, Deltacommissie, The Hague

Den Hengst, S. 2012. The deltadike concept and the effectiveness of various dike reinforcement alternatives. MSc thesis, Delft University of Technology, Delft.

Denzin, N.K. 1978. The Research Act. McGraw-Hill, New York, USA.

Dietz, T., Ostrom, E., & Stern, P.C. 2003. The struggle to govern the commons. Science, 302, 1907-1912.

Dovers, S. 2005. Clarifying the imperative of integration research for sustainable environmental management. Journal of Research Practice 1 (2), Article M1. Retrieved from http://jrp.icaap.org/index.php/jrp/article/view/11/30.

DPRD, Deltaprogramma Rijnmond-Drechtsteden 2013. Concept Gebiedsrapportage Eiland van Dordrecht.

Dudley E., Ashley, R., Manojlovic, N., van Herk, S., Blanksby J. 2013. Learning and Action Alliances for innovation and active learning in a European context. International Conference on Flood Resilience. Exeter, UK.

EA, 2000. department of the environment, transport and the regions, environment agency and institute of environmental health. Guidelines for Environmental Risk Assessment and Management, 2nd edition.The Stationery Office, London.

Ebregt, J., Eijgenraam, C. & Stolwijk, H. 2005. Kosten-baten analyse voor Ruimte voor de Rivier, deel 2 - Kosteneffectiviteit van maatregelen en pakketten. Centraal Plan Bureau, The Hague.

EC, European Commission 2004. Flood risk management. Flood prevention, protection and mitigation, Communication from the Commission to the Council, The European parliament, The European Economic and Social Committee of the Regions, Brussels, 12.07.2004 COM 472 final 2006/0005(COD), http://ec.europa.eu/environment/water/ flood risk/.

EC, European Commission, 2006. Directive of the European Parliament and of the Council on the assessment and management of floods (Proposal), {SEC(2006) 66} COM(2006) 15 final 2006/0005(COD), http://ec.europa.eu/environment/water/ flood risk/.

EC, European Commission (2008): Identification and designation of European critical infrastructures and the assessment of the need to improve their protection. Council Directive 2008/114/EC.

ECTP, 2005. Strategic Research Agenda for the European Construction Sector; Achieving a sustainable and competitive construction sector by 2030. Available at www.ectp.org, last accessed 29 January 2011.

Edelenbos, J., & Klijn, E.H. 2006. Managing stakeholder involvement in decision making: A comparative analysis of six interactive processes in the Netherlands. Journal of public administration research and theory 16, 417-446.

Edelenbos, J., & Teisman, G. 2013. Water governance capacity: The art of dealing with a multiplicity of levels, sectors and domains. International Journal of Water Governance 1, 89–108 89 DOI: 10.7564/12-IJWG5

Edelenbos, J., van Buuren, A., & van Schie, N. 2011. Co-producing knowledge: joint knowledge production between experts, bureaucrats and stakeholders in Dutch water management projects. Environmental Science & Policy 14, 675-684.

Ehrmann, J.R. & Stinson, B.L. 1999. Joint fact-finding and the use of technical experts, in: Susskind, L., S. McKearnan and J. Thomas-Larmer (eds.), The consensus building handbook. A comprehensive guide to reaching agreement, London: Sage, 375-400.

Eisenhardt, K.M. 1989. Building Theories from Case Study Research, in: The Academy of Management Review 14, 532-550.

ERA-Net CRUE. 2007. National research programmes on flood risk management Across Europe. CRUE Report Work Package 2. Federal Ministry of Agriculture, Forestry, Environment and Water Management, Vienna, Austria.

Evans, E.P., Ashley R.M., Hall J., Penning-Rowsell, E., Sayers, P., Thorne C. & Watkinson, A. 2004. Foresight. Future Flooding Vol II – Managing future risks. UK Office of Science and Technology. April, London, UK.

Farrelly, M., & Brown, R. 2011. Rethinking urban water management: Experimentation as a way forward? Global Environmental Change 21, 721-732.

Farrelly, M., Brown, R.R. and Davis, C. 2009. Can demonstration projects act as a mechanism for promoting a transition? Paper Submitted to the 6th International Water Sensitive Urban Design Conference and 3rd Hydropolis, Perth, Western Australia 6-8th May, 2009.

Farrelly, M.A. & Brown, R.R., 2008. Professional perceptions on institutional drivers and barriers to advancing diverse water options in Australia. Proc. 11th Int. Conf. Urban Drainage. Edinburgh.

Fischer, F. 1990. Technocracy and the politics of expertise, New York: Sage.

Flikweert, J. & Simm, J. 2008. Improving performance targets for flood defence assets. Journal of Flood Risk Management 1, 201-212.

Flood, R. & L. 1998. Action research and the management and systems sciences. Systemic Practice and Action Research 11, 79–101.

Flood, R.L. & Romm, N.R.A., 1996. Contours of diversity management and triple loop learning. Kybernetes 25, 154–163.

FloodProBE 2013, Technologies for the cost-effective Flood Protection of the Built Environment, www.floodprobe.eu [accessed 16/05/2013]

Flyvbjerg, B. 2006. Five misunderstandings about Case-Study research, Qualitative Inquiry, 12, 219-245.

Flyvbjerg, B. 2007. Truth and Lies About Megaprojects, Faculty of Technology, Policy, and Management, Delft University of Technology, The Netherlands.

Folke, C. 2006. Resilience: the emergence of a perspective for social-ecological systems analysis. Global Environmental Change 16, 253-267.

Folke, C., Hahn, T., Olsson, P. & Norberg, J. 2005. Adaptive governance of social-ecological systems. Annual Review of Environment and Resources 30, 441-473.

Franke, U.J. 1999. The virtual web as a new entrepreneurial approach to network organizations. Entrepreneurship and Regional Development 11, 203-230.

Geels 2005. Technological transitions and system innovations, A Co-Evolutionary and Socio-Technical Analysis, Edward Elgar, Cheltenham, UK; Northampton, MA, USA, p. 279.

Geels, F. & Kemp, R. 2000. Transities vanuit sociotechnisch perspectief, International Centre for Integrative Studies, Maastricht, The Netherlands.

Geels, F.W. & Schot, J. 2007. Typology of transition pathways in socio-technical systems. Research Policy 36, 399–417.

Geldof, G.D. 2007. The Three Points Approach. Proceedings 5th South Pacific Stormwater Conference, Auckland, May 2007.

Gentleman, A. 2007. Architects aren't ready for an urbanised planet. The International Herald Tribune, 11–29.

George, A. L. & Bennett, A. 2004. Case studies and theory development in social science. Cambridge, MA: MIT Press.

Gersonius, B., Ashley, R., Pathirana, A. & Zevenbergen, C. 2010. Managing the flooding system's resiliency to climate change. Eng Sustain 163, 15–22.

Gersonius, B. 2012. The resilience approach to climate adaptation applied for flood risk. UNESCO-IHE PhD thesis, CRC Press, Leiden. ISBN 978-0-415-62485-5

Gersonius, B., Ashley R. & Zevenbergen, C. 2012. The identity approach for assessing socio-technical resilience to climate change: example of flood risk management for the Island of Dordrecht, Natural Hazards and Earth System sciences, 12, 1-8.

Gersonius, B., Veerbeek, W., Subhan, A., Zevenbergen, C. 2010. Toward a more flood resilient urban environment: the Dutch Multi-level Safety approach to flood risk management. In: Resilient Cities: Cities and Adaptation to Climate Change – Proceedings of the Global Forum 2010, Bonn, Germany. Otto-Zimmerman (ed.) Springer. DOI 10.1007/978-94-007-0785-6_28.

Gersonius, B., Zevenbergen, C. & Herk, van S. 2007. Managing flood risk in the urban environment: linking spatial planning, risk assessment, communication and policy. In: Adaptive and Integrated Water Management: coping with complexity and uncertainty. C Pahl Wostl, P Kabat, and J Möltgen (Eds.). Springer ISBN 978 3 540 75940 9.

Gibbons, M.N. 2001. The potential of transdisciplinarity. Birkhauser Verlag, Basel.

Gill, S. E., Handley, J. F., Ennos, A. R., & Pauleit, S. 2007. Adapting cities for climate change: the role of the green infrastructure. Built environment 33, 115-133.

Global Water Partnership – Technical Advisory Committee (GWP-TAC). 2004. Integrated water resources management (IWRM) and water efficiency plans by 2005. Why, what and how? TAC Background Papers No. 10. GWP, Stockholm, Sweden.

Graddy, E. A., & Chen, B. 2006. Influences on the size and scope of networks for social service delivery. Journal of Public Administration Research and Theory 16, 533-552.

Green, C. 2010. Towards sustainable flood risk management. International Journal of Disaster Risk Science, 1, 33-43.

Gunderson, L.H., Carpenter, S.R., Folke, C., Olsson, P. & Peterson, G. 2006 Water RATs (Resilience, Adaptability, and Transformability) in Lake and Wetland Socio-Ecological Systems" Ecology and Society, 11 (1), 16 (online).

Haasnoot, M. 2013. Anticipating change: sustainable water policy pathways for an uncertain future.

Hajer, M., & Zonneveld, W. 2000. Spatial planning in the network society-rethinking the principles of planning in the Netherlands. European Planning Studies 8, 337-355.

Hajer, M. 2003. Policy without polity? Policy analysis and the institutional void. Policy sciences 36, 175-195.

Hajer, M., Sijmons, D. & Feddes, F. 2006 Een plan dat werkt. Ontwerp en politiek in de regionale planvorming, Nai Uitgevers, Rotterdam.

Hamin, E M., & Gurran N. 2009. Urban form and climate change: Balancing adaptation and mitigation in the US and Australia. Habitat International 33, 238-245.

Hanf, K. I., & Scharpf F.W. 1978. Interorganizational policy making: Limits to coordination and central control. London: Sage.

Hargrove, R. 2002. Masterful Coaching. Revised Edition. Jossey-Bass/Pfeiffer, Wiley, USA.

Healey P. 1998, Institutionalist theory, social exclusion and governance, in Social Exclusion in European Cities, Eds A Madanipour, G Cars, J Allen (Jessica Kingsley, London) 53 - 73

Hegger, D.L.T., Lamers, M., Van Zeijl-Rozema, A., Dieperink, C., 2012. Conceptualising joint knowledge production in regional climate change adaptation projects: success conditions and levers for action. Environmental Science and Policy 15, 52-65.

Hertogh, M, Baker, S., Staal-Ong, P.L., Westerveld, E. 2008 Managing large infrastructure projects; research based on Best Practices and Lessons Learnt In Large Infrastructure Projects in Europe. 978-90-810025-2-3

Hertogh, M.J.C.M., Westerveld, E., 2010 Playing with Complexity. Management and organisation of large infrastructure projects, PhD thesis. Erasmus University Rotterdam.

Hillier, J. 2000. Going round the back? Complex networks and informal action in local planning processes. Environment and Planning A 32, 33-54.

Huitema, D., & Meijerink, S. 2010. Realizing water transitions: the role of policy entrepreneurs in water policy change. Ecology and Society 15, 26.

Huitema, D., Mostert, E., Egas, W., Moellenkamp, S., Pahl-Wostl, C., & Yalcin, R. 2009. Adaptive water governance: assessing the institutional prescriptions of adaptive (co-) management from a governance perspective and defining a research agenda. Ecology and Society, 14, 26.

Hulsker, W., Wienhoven, M., van Diest, M. & Buijs, S., 2011. Evaluatie ontwerpprocessen Ruimte voor de Rivier. Ecorys.

Hult, M. & Lennung, S. 1980, Towards a definition of action research: a note and a bibliography", Journal of Management Studies 17, 241-50.

Huntjens, P., Pahl-Wostl, C., Rihoux, B., Schlüter, M., Flachner, Z., Neto, S., Koskova, R., Dickens, C. & Nabide Kiti, I. 2011. Adaptive water management and policy learning in a changing climate: a formal comparative analysis of eight water management regimes in Europe, Africa and Asia. Environmental Policy and Governance 21, 145-163.

I&M. 2011. Main report of the 2nd Flood Protection Programme. Basisrapportage van het 2e Hoogwaterbeschermingsprogramma (HWBP-2). (In Dutch).

Intergovernmental Panel on Climate Change (IPCC), Climate Change 2007 (IPCC, Geneva, 2007) http://www.ipcc.ch/.

Ison, R., Watson, D., 2007. Illuminating the possibilities for social learning in the management of Scotland's water. Ecology and Society 12 (1) (online) URL: <http:www.ecologyandsociety.org/vol12/iss1/art21/>

Ison, R., Blackmore, C., & Iaquinto, B. L. 2013. Towards systemic and adaptive governance: Exploring the revealing and concealing aspects of contemporary social-learning metaphors. Ecological Economics 87, 34-42.

Jeffrey, P. and Seaton, R.A.F. 2003/2004. A conceptual model of 'Receptivity' applied to the design and deployment of water policy mechanisms. Environmental Sciences 1, 277-300.

Jeffrey, P., & Gearey, M. 2006. Integrated water resources management: lost on the road from ambition to realisation?

Jentoft, S. 2007. Limits of governability: Institutional implications for fisheries and coastal governance. Marine Policy, 31, 360-370.

Jick, T.D. 1979. Mixing Qualitative and Quantitative Methods: Triangulation in Action. Administrative Science Quarterly 24, 602-611.

Jiggins, J., Van Slobbe, E., & Röling N. 2007. The Organisation of Social Learning in Response to Perceptions of Crisis in the Water Sector of the Netherlands. Environmental Science & Policy 10, 526-536.

Jonkman, S.N. 2007. Loss of life estimation in flood risk assessment, theory and applications. Dissertation Delft University of Technology, Delft

Jonkman, S.N., Stive, M.J.F. & Vrijling, J.K. 2005. New Orleans is a lesson to the Dutch. J Coast Res 21, xi–xii.

Kabat, P., Fresco, L.O., Stive, M.J.F., Veerman, C.P., van Alphen, J.S.L.J., Parmet, B.W.A.H., Hazeleger, W. & Katsman, C.A. 2009. Dutch coasts in transition. Nature Geoscience 2, 450-452.

Kates, R.W., Colten, C.E., Laska, S. & Leatherman, S.P. 2006. Reconstruction of New Orleans after Hurricane Katrina: a research perspective. Proceedings of the National Academy of Sciences 103, 14653–14660

Katsman, C.A., Sterl, A., Beersma, J.J., van den Brink, H.W., Hazeleger, W. et al. 2011. Exploring high-end scenarios for local sea level rise to develop flood protection strategies for a low-lying delta – the Netherlands as an example. Clim Change. doi:10.1007/s10584-011-0037-5. Accessed 30 Jan 2012

Katz, J. S., & Martin, B. R. 1997. What is research collaboration? Research Policy, 26, 1-18.

Keen, M., Brown, V., Dybal, R., 2005. Social Learning in Environmental Management: Towards a sustainable future. Earthscan, London, UK.

Kemp, R., Schot, J. & Hoogma, R., 1998. Regime shifts to sustainability through processes of niche formation: the approach of strategic niche management. Technology Analysis & Strategic Management 10, 175-198.

Kerkhof, M. van de, Wieczorek, A., 2005. Learning and stakeholder participation in transition processes toward sustainability: methodological considerations. Technological Forecasting and Social Change 72, 733-747.

Kern, K., Alber, G. & Policy, C. 2008. Governing climate change in cities: modes of urban climate governance in multi-level systems. In: Competitive Cities and Climate Change, OECD conference proceedings 171-196. Milan, Italy, 9-10 October 2008.

Kickert, W. J. M., Klijn, E.H. & Koppenjan, J. F. M. eds. 1997. Managing complex networks: Strategies for the public sector. London: Sage.

Kilduff, Martin, & Wenpin Tsai. 2003. Social Networks and Organizations. London: Sage Press.

King, C. & Jiggins, J., 2002. A systematic model and theory for facilitating social learning. In: Leeuwis, C., Pyburn, R. (Eds.), Wheelbarrows Full of Frogs. Koninklijke Van Gorcum, Assen, The Netherlands, 85–105.

Kingdon, J.W. 1984. Agendas, alternatives, and public policies. Boston: Little, Brown and Company.

Kjær, A. 2004. Governance, Cambridge: Polity.

Klein, J.T. 2001. Transdisciplinarity: joint problem solving among science, technology, and society: an effective way for managing complexity. Basel: Birkhauser Verlag.

Klijn E. & van Twist, M.J.W. 2008. Zicht op de omgeving: Een netwerkbenadering om de omgeving te analyseren. In: A. Edwards & L. Schaap (redactie), Vaardigheden voor de publieke sector (pp. 37-61). Bussum, Coutinho.

Klijn, E. H., Edelenbos, J., & Steijn, B. 2010. Trust in Governance Networks Its Impacts on Outcomes. Administration & Society, 42, 193-221.

Klijn, F., Samuels, P., & Van Os, A. 2008. Towards flood risk management in the EU: State of affairs with examples from various European countries. International Journal of River Basin Management, 6, 307-321.

Klijn, F., de Bruin, D., de Hoog, M. C., Jansen, S., & Sijmons, D. F. 2013. Design quality of Room-for-the-River measures in the Netherlands: role and assessment of the Quality Team (Q-team). International Journal of River Basin Management, (just-accepted), 1-23.

Klinke, A., and Renn, O. 2006. Systematic risks as a challenge for policy making in risk governance. Forum: Qualitative Social Research 1, Art 33

Knight, L. 2002. Network learning: exploring learning by interorganizational networks. Human Relations, 55, 427-454.

Knight. L, & Pye, A. 2004. Exploring the Relationships between Network Change and Network Learning. Management learning 4, 473-490. doi: 10.1177/1350507604048274

Kolen, B., & Kok, M. 2011. Optimal investment in emergency management in a multi layer flood risk framework. In 5th International Conference on Flood Management Tokyo (Japan).

Kolen, B., Maaskant, B., Hoss, F. 2010. Meerlaagsveiligheid: Zonder normen geen kans, In: Ruimtelijke Veiligheid en risicobeleid, Nummer 2.

Kooiman, J. 1993. Modern governance: new government-society interactions. Sage Publications Ltd, London, UK.

Kooiman, J., 2003. Governing as Governance. Sage, London.

Kooiman, J., Bavinck, M., Jentoft, S., Pulin, R. (Eds.), 2005. Fish for Life: Interactive Governance for Fisheries. Amsterdam University Press, Amsterdam.

Koppenjan, J. F. M. & Klijn, E.-H. 2004. Managing Uncertainties in Networks: A Network Approach to Problem Solving and Decision Making, London: Routledge.

Koukoui, N., Gersonius, B., Schot, P., Van Herk, S. 2013. Adaptation Tipping Points and Opportunities for the Management of Urban Flood Risk - Case Study in Dordrecht, the Netherlands. European Climate Change Adaptation Conference (ECCA), Hamburg, Germany.

Kron, W. 2009. Flood insurance: from clients to global financial markets. J. Flood Risk Management 2, 68-75.

Levinthal, D.A. 1998. The slow pace of rapid technological change: gradualism and punctuation in technological change, Industrial and Corporate Change 7, 32.

Lincoln, Y. S. 1989. Fourth generation evaluation. Sage.

Lindblom, C. E. & D.K. Cohen, 1979, Usable Knowledge: Social Science and Social Problem Solving, New Haven, CT: Yale University Press.

Lipnack, J.H. & Stamps, J. 1997. Virtual Teams, reach across space, time and organizations with technology, John Wiley, New York

Lips, N. (2012). Application of storylines for strategy development in flood risk management; A case study on the flooding of the island of Dordrecht, the Netherlands. MSc thesis University of Utrecht.

LMW 2010. Leven met Water, www.levenmetwater.nl [accessed 14/09/09]

Loorbach, D. 2010. Transition Management for Sustainable Development: A Prescriptive, Complexity Based Governance Framework. Governance, 23, 161-183

Loorbach, D. 2007. Transition management. New mode of governance for sustainable development. International Books, Utrecht, The Netherlands.

Lundqvist, L. 2004. Integrating Swedish Water Resource Management: a multi-level governance trilemma. Local Environment 9, 413-424.

Manring, S.L., Rivkin, G.W., Rivkin, M.D. 2003. The application of network and learning organization models to improve ecosystem management: Monroe 2020 case study. Environmental Practice 2, 13-27.

MARE 2011, Managing Adaptive REsponses to changing flood risk, www.mare-project.eu [accessed 10/05/2011]

Marsh, D., & R. A. W. Rhodes, eds. 1992. Policy networks in British government. Oxford: Clarendon Press.

Mauch, C. 2009. in Natural disasters, cultural responses. C. Mauch and C. Pfister, Eds.Lexington Books, chap. 1. pp. 1-17.

Maylor, H., Brady, T., Cooke-Davies, T. & Hodgson, D. 2006. From projectification to programmification. International Journal of Project Management 24, 663-674.

McKay, J., & Marshall, P. 2001. The dual imperatives of action research. Information Technology & People, 14, 46-59.

Medema, W., McIntosh, B. S., Jeffrey, P.J. 2008 From Premise to Practice: a Critical Assessment of Integrated Water Resources Management and Adaptive Management Approaches in the Water Sector. Ecology & Society 13, 29.

Merz, B., Hall, J., Disse, M., & Schumann, A. 2010. Fluvial flood risk management in a changing world. Natural Hazards and Earth System Science, 10, 509-527.

Meuleman L. 2008 Public Management and the Metagovernance of Hierarchies, Networks and Markets: The feasibility of Designing and Managing Governance style combinations. Physica-Verlag, Heidelberg, Germany.

Miles, R.E. & Snow, C.C. 1986. Network organisations: new concepts for new forms. California Management Review 28, 62-73.

Milly P. C. D., Betancourt J., Falkenmark M., Hirsch R M., Kundzewicz Z W., Lettenmaier D P., & Stouffer R J. 2008, Stationarity Is Dead: Whither Water Management?, in: Science 319, 5863(573-574)

Ministry of Transport, Public Works and Water Management 2000. Anders omgaan met water. Waterbeleid voor de 21e eeuw - Kabinetsstandpunt. The Hague, the Netherlands: Ministerie van Verkeer en Waterstaat.

Ministry of Infrastructure & Envrionment 2011. Syntheserapport Gebeidspilots Meerlaagsveiligheid. In Dutch.

Mintzberg, H., D. Raisinghani & A. Therot 1976. The structure of unstructured decision processes, Administrative Science Quarterly 21, 246-75.

Moellenkamp, S., Lamers, M., Huesmann, C., Rotter, S., Pahl-Wostl, C., Speil, K. & Phol, W. 2010. Informal participatory platforms for adaptive management: insights into niche-finding, collaborative design and outcomes from a participatory process in the Rhine Basin. Ecol. Soc. 15, 41

Mommaas, H., Janssen, J. 2008. Towards a synergy between 'content' and 'process' in Dutch spatial planning: the Heuvelland case. Journal of Housing and the Built Environment 23, 21–35.

Mostert, E., Pahl-Wostl, C., Rees, Y., Searle, B., Tàbara, D., & Tippett, J. 2007. Social learning in European river-basin management: barriers and fostering mechanisms from 10 river basins. Ecology and society, 12, 19.

Munich Re. 2009. Topics geo, Natural Catastrophes 2008: Analyses, assessments, positions. Munich: Munich Reinsurance Group.

Muro, M., Jeffrey, P., 2008. A critical review of the theory and application of social learning in participatory natural resource management processes. Journal of Environmental Planning and Management 51, 325–344.

National Committee on Levee Safety, Recommendations for a National Levee Safety Program: A Report to Congress from the National Committee on Levee Safety, (National Committee on Levee Safety, 15 January 2009 (Draft)).

Neufville, R. de, 2004. Uncertainty Management for Engineering Systems Planning and Design, Engineering Systems Symposium, MIT

Neuvel, J. M., & van den Brink, A. 2009. Flood risk management in Dutch local spatial planning practices. Journal of Environmental Planning and Management, 52, 865-880.

Newman R., Ashley R M., Molyneux-Hodgson S., Cashman A. 2011. Managing water as a socio-technical system: the shift from 'experts' to 'alliances'. Proc. Institution of Civil Engineers. Engineering Sustainability 164, 95-102.

Nokes, S. 2007. The Definitive Guide to Project Management. 2nd Ed.n. London, Financial Times / Prentice Hall. ISBN 978-0-273-71097-4

Nonaka, I., & H. Takeuchi. 1995. The knowledge-creating company. Oxford University Press, Oxford, UK.

OECD 2010. Étude de l'OCDE sur la gestion des risques d'inondation: Bassin de la Loire, France 2010, ISBN:9789264017740

Olsson, P., L. H. Gunderson, S. R. Carpenter, P. Ryan, L. Lebel, C. Folke, and C. S. Holling. 2006. Shooting the rapids: navigating transformations to adaptive governance of social-ecological systems. Ecology and Society 11, 18 [online] URL: http:// www.ecologyandsociety.org/vol11/iss1/art18/.

Opperman, J.J., Galloway, G.E., Fargione, J., Mount, J.F., Richter, B.D. & Secchi, S. 2009. Sustainable floodplains through large-scale reconnection to rivers. Science 326, 1487.

Pahl-Wostl C., 2002. Towards sustainability in the water sector: the importance of human actors and processes of social learning. Aquatic Sciences, 64, 394-411.

Pahl-Wostl C., ed. 2006. Framework for adaptive water Management regimes and for the transition between regimes. NeWater project, Report Series No. 12.

Pahl-Wostl, C. 2006. The importance of social learning in restoring the multifunctionality of rivers and floodplains. Ecology and society, 11, 10.

Pahl-Wostl C., Holtz G., Kastens B. & Knieper C. 2010. Analyzing complex water governance regimes: the Management and Transition Framework. Environmental Science & Policy, 13, 571-581.

Pahl-Wostl, C. 2007. Transitions towards adaptive management of water facing climate and global change. Water Resources Management, 21, 49-62.

Pahl-Wostl, C. 2009. A conceptual framework for analysing adaptive capacity and multi-level learning processes in resource governance regimes. Global Environmental Change, 19, 354-365.

Pahl-Wostl, C., Craps, M., Dewulf A., Mostert, E., Tabara D. and Taillieu T. 2007. Social learning and water resources management. Ecology and Society, 12(2): 5. (online) URL: http://www.ecologyandsociety.org/vol12/iss2/art5/.

Pahl-Wostl, C., Lebel, L., Knieper, C., & Nikitina, E. 2012. From applying panaceas to mastering complexity: Toward adaptive water governance in river basins. Environmental Science & Policy 23, 24-34.

Patton, M. Q. 2008. Utilization-focused evaluation. Sage.

Patton, M. Q. 2011. Developmental evaluation: Applying complexity concepts to enhance innovation and use. Guilford Press.

PDR 2011a. 18e Voortgangsrapportage Ruimte voor de Rivier 1 januari 2011 - 30 juni 2011.

PDR 2011b. 19e Voortgangsrapportage Ruimte voor de Rivier 1 juli 2011 - 30 december 2011.

Pearce, L. 2003. Disaster management and community planning, and public participation: how to achieve sustainable hazard mitigation. Natural Hazards 28, 211-228.

Petts, J. & C. Brooks, 2006. Expert conceptualizations of the role of lay knowledge in environmental decision making: challenges for deliberative democracy, in: Environment and Planning A 38, 1045-1059.

Pielke, R.A. 2007. The honest broker: making sense of science in policy and politics. Cambridge, New York: Cambridge University Press.

Pierre, J., & Peters, B. 2000. Governance, politics and the state. St. Martin's Press, New York, USA.

Pieterse, N., Knoop J., Nabielek, K., Pols L. & Tennekes, J. 2009. Overstromingsrisicozonering in Nederland. Hoe in de ruimtelijke ordening met overstromingsrisico's kan worden omgegaan. PBL, Den Haag, Bilthoven.

PKB, 2006. Planologische kernbeslissing: Room for the River (key Decision on Spatial Planning: Room for the Rhine Branches). Room for the River, Project Office (in Dutch).

Plummer, R., FitzGibbon, J., 2004. Some observations on the terminology in co-operative environmental management. Journal of Environmental Management 70, 63–72.

Potter, K., Ward, S., Shaw, D., Macdonald, N., White, I., Fischer, T. , Butler, D. & Kellagher, R.B.B. 2011. Engineers and planners: sustainable water management alliances, Engineering Sustainability 164, 239-247.

Priemus, H. 2007. The Network Approach: Dutch Spatial Planning between Substratum and Infrastructure Networks, European Planning Studies 15, 667 – 686.

Project Management Institute 2008. The standard for programme management, Second Edition ed. PMI, Newton Square, PA, USA.

Provan, K. G., & Kenis, P. 2008. Modes of network governance: Structure, management, and effectiveness. Journal of public administration research and theory, 18, 229-252.

Provan, K.G., Keith G., Fish, A. & Sydow J. 2007. Interorganizational networks at the network level: A review of the empirical literature on whole networks. Journal of Management 33, 479–516.

Provincie Overijssel (2007) Handreiking Ruimtelijke Kwaliteit IJssel

PZH, 2011. Third Round Statutory Assessment Primary Flood Defences, Dike Ring Area 22 - Island of Dordrecht. Derde Ronde Toetsing Primaire Waterkeringen, Dijkringgebied 22 - Eiland van Dordrecht. (In Dutch).

Ratcliffe, J., Stubbs, M. & Shepherd, M. 2004. Urban Planning and Real Estate Development (London, Spon Press).

Raven, R., Van den Bosch, S., & Weterings, R. 2010. Transitions and strategic niche management: towards a competence kit for practitioners. International Journal of Technology Management, 51, 57-74.

Renn, O., 1995. Style of using scientific expertise: a comparative framework, in: Science and Public Policy 22, 1-11.

Reuber, J., Schielen, R. & Barneveld, H. J. 2005. Preparing a river for the future – the River Meuse in the year 2050 Proceedings of the 3rd International Symposium on Flood Defence 25–27 May 2005 Nijmegen, The Netherlands

Rhodes, R. A. W. 1997. Understanding governance. Buckingham, UK: Open University Press.

Rhodes, R. 1996. The New Governance: Governing without Government. Political studies 44, 652-667.

Rijke J. S., De Graaf R. E., Van de Ven F. H. M., Brown R. R. & Biron D. J. 2008. Comparative case studies towards mainstreaming water sensitive urban design in Australia and the Netherlands. In: Proceedings of the 11th International Conference on Urban Drainage (ICUD), Edinburgh, Scotland, 31 August - 5 September 2008.

Rijke, J., Brown, R., Zevenbergen, C., Ashley, R., Farrelly, M., van Herk, S. & Morison, P. 2012a. Fit-for-purpose governance: A framework to operationalise adaptive governance. Environmental Science & Policy 22, 73-84.

Rijke, J., van Herk, S., Zevenbergen, C., Ashley, R. 2012b. Room for the River: Delivering integrated river basin management in the Netherlands. International Journal of River Basin Management 10, 1-21.

Rijke, J., Farrelly, M., Brown, R., Zevenbergen, C. 2013. Configuring transformative governance to enhance resilient urban water systems. Environmental Science and Policy 25, 62-72.

Rijke, J., van Herk, S., Zevenbergen, C., Ashley, R., Hertogh, M. & Ten Heuvelhof, E. (in press). Adaptive programme management through a balanced performance/strategy oriented focus across all management levels. In press at: International Journal of Project Management

Rijke, J., Brown, R., van Herk, S. & Zevenbergen, C. (under review) Governance for strategic planning and delivery of adaptation. Global Environmental Change.

Rip, A., & Kemp, R. P. M. 1998. Technological Change. In: Rayner S., Malone EL (editors) 327-399. Battelle Press.

Ritson, G., Johansen, E. & Osborne, A. 2011. Successful programs wanted: Exploring the impact of alignment. Project Management Journal 43, 21-36.

Rittel, H. & Webber, M. 1973, 'Dilemmas in a general theory of planning', Policy Sciences, 4, 155–169.

Robson, C. 2002. Real World Research. Blackwell, Oxford.

Rose, A.Z. 2004. Economic Principles, Issues, and Research Priorities in Natural Hazard Loss Estimation, In: Modeling the Spatial Economic Impacts of Natural Hazards, Eds. Y. Okuyama and S. Chang, Springer, 13 – 36

Rotmans J., Kemp R. and Van Asselt M. 2001. Transition Management in Public Policy, Foresight 3, 15-31.

Rotmans, J. 2005, Societal innovation: Between dream and reality lies complexity, Inaugural speech, Erasmus University Rotterdam.

Loorbach, D. & Rotmans, J. 2006. Managing transitions for sustainable development, 187-206. Springer Netherlands.

Rotmans, J. & D. Loorbach 2009. Complexity and Transition Management. Journal of Industrial Ecology 13, 184-196.

Ruitenbeek, H.J., & Cartier, C.M. 2001. The invisible wand: adaptive co-management as an emergent strategy in complex bio-economic systems (Vol. 34). Bogor, Indonesia: Center for International Forestry Research.

RvR Raad voor Vastgoed Rijksoverheid 2008. Richtlijn onteigening Ruimte voor de Rivier

RWS Rijkswaterstaat 2010. Werken met de Waterwet; juridische leidraad voor Rijkswaterstaat.

Saeijs, H. 1991. Integrated water management: a new concept. From treating of symptoms towards a controlled ecosystem management in the Dutch Delta. Landscape and Urban Planning 20, 245-255.

Salet, W. et al. (eds.) 2003. Metropolitan Governance and Spatial Planning – Comparative Case Studies of European City Regions. London: Spon Press.

Sandström, U.F., 2002. Green Infrastructure planning in urban Sweden. Planning Practice and Research,17, 373-385.

Scharpf, F. W. 1997. Games real actors play: Actor-centered institutionalism in policy research. Boulder, CO: Westview Press.

Schlager, E. 1999. A comparison of frameworks, theories, and models of policy processes. Pages 233–260 in P. A. Sabatier, editor. Theories of the policy process. Westview Press, Boulder, Colorado, USA.

Scholz, R.W. 2000. Mutual learning sessions. Zurich: Haffmanns.

Scholz, R.W., & Tietje, O. 2002. Embedded case study methods: integrating quantitative and qualitative knowledge. Thousand Oaks, California: Sage Publications.

Schön, D.A. 1983. The reflective practitioner: how professionals think in action, Basic Books. ISBN 0465068782

Schön, D.A. & Rein, M. 1994. Frame reflection: Toward the resolution of intractable policy controversies. New York: Basic Books.

Schramm, W. & Porter, W.E. 1982, Men, women, messages, and media: understanding human communication. 2nd ed. New York: Harper and Row.

Schut, M., Leeuwis, C. van Paassen, A. 2010. Room for the River: Room for Research? The case of depoldering De Noordwaard, the Netherlands. Science and Public Policy,37, 611-627.

Scott M., White I., Kuhlicke C. et al 2013. Living with Flood risk. Planning theory and practice 14, 103-140.

Sellers, J.M. 2002. Governing from Below, Urban Regions and the Global Economy. Cambridge University Press, Cambridge.

Senge, P. M. 1990. The fifth discipline: the art and practice of the learning organisation. Currency Doubleday, New York, New York, USA.

Shehu, Z., Akintoye, A. 2009. The critical success factors for effective programme management: a pragmatic approach. The Built & Human Environment Review 2, 1-24.

Smith, A., & Stirling, A. 2010. The politics of social-ecological resilience and sustainable socio-technical transitions. Ecology and Society, 15, 11.

Spit, T.H.M., & P.R. Zoete 2006. Ruimtelijke ordening in Nederland: een wetenschappelijke introductie in het vakgebied. The Hague: Sdu.

Stalenberg, B., Vreeker, R., Edelenbos, J., Verweij, S., Winsemius H.C., Van Herk, S., Smid, G.J.P., Van Middendorp, B. 2011. Klimaat in Ruimtelijke Keuzes; het dialoogondersteunend afwegingskader (DAK); theorie en praktijk.

Steyaert, P. & Jiggins, J. 2007. Governance of complex environmental situations through social learning: a synthesis of SLIM's lessons for research, policy and practice. Environ. Sci. Policy 10, 575-586.

TASKFORCE HWBP, 2012. Een dijk van een programma - Naar een nieuwe aanpak van het Hoogwaterbeschermingsprogramma.

Taylor, A.C. 2008. Leadership in Sustainable Urban Water Management: An Investigation of the Champion Phenomenon within Australian Water Agencies. Report No. 08/01, National Urban Water Governance Program, Monash University, ISBN: 978-0-9804298-5-5.

Teisman, G.R. 2000. Models for research into decision-making processes: on phases, streams and decision-making rounds, in: Public Administration, 78, 4, 937-956.

Termeer, C.J.A.M., Koppenjan, J.F.M. 1997. Managing perceptions in networks. In WJM Kickert, E.-H. Klijn, & JFM Koppenjan (Eds.), Managing complex networks: Strategies for the public sector, 79-97. London: Sage Publications.

Therivel, R. (2012). Strategic environmental assessment in action. Routledge.

Thomas, J. S., & Durham, B. 2003. Integrated Water Resource Management: looking at the whole picture. Desalination, 156, 21-28.

Toonen, T. A. 1990. The Unitary State as a System of Co-governance: The Case of The Netherlands. Public Administration, 68, 281-296.

Treby, E. J., Clark, M. J., & Priest, S. J. 2006. Confronting flood risk: implications for insurance and risk transfer. Journal of environmental management, 81, 351-359.

Tuinstra, W. 2008. Learning and evaluation in Integrated Sustainability Assessment, Int. J. Innovation and Sustainable development, Vol. 3, Nos. 1/2.

Tzoulas, K., Korpela, K., Venn, S., Yli-Pelkonen, V., Kaźmierczak, A., Niemela, J., & James, P. 2007. Promoting ecosystem and human health in urban areas using Green Infrastructure: A literature review. Landscape and urban planning 81, 167-178.

UFM 2008. Urban Flood Management Dordrecht, www.ufmdordrecht.nl [accessed 02/04/08]

UNISDR 2012. Towards a Post-2015 Framework for Disaster Risk Reduction. http://www.unisdr.org/files/25129_towardsapost2015frameworkfordisaste.pdf

UNISDR 2011a. Global Assessment Report on Disaster Risk Reduction. Revealing Risk, Redefining Development. United Nations International Strategy for Disaster Reduction Secretariat, Geneva.

UNISDR 2011b. HYOGO FRAMEWORK FOR ACTION 2005-2015. Building the Resilience of Nations and Communities to Disasters: a mid-term review. United Nations, Geneva, Switzerland.

V&W 2006. Ministerie van V&W, Ministerie van VROM en Ministerie van LNV Planologische Kernbeslissing Ruimte voor de Rivier deel 4, Vastgesteld besluit, Nota van Toelichting

V&W, Ministry of Transport, Public Works and Water Management 2006. Beleidslijn Grote Rivieren, The Hague, the Netherlands: Ministerie van Verkeer en Waterstaat. in Dutch

V&W, Ministry of Transport, Public Works and Water Management 2008. Nationaal Water Plan 2009-2015. The Hague, the Netherlands: Ministerie van Verkeer en Waterstaat. in Dutch with summary in English

Van Buuren & Nooteboom 2009. Evaluating strategic environmental assessment in The Netherlands: content, process and procedure as indissoluble criteria for effectiveness. Impact Assessment and Project Appraisal 27 (2): 145-154.

Van Buuren, A. 2006. Competente besluitvorming: het management van meervoudige kennis in ruimtelijke ontwikkelingsprocessen. Centraal Boekhuis. ISBN 9059314794.

Van de Ven, F., Gersonius, B., Graaf, R., Luijendijk, E. & Zevenbergen, C. 2011. Creating water robust urban environments in the. Journal of Flood Risk Management, 4, 273-280.

Van de Ven, F., Luyendijk, E., Gunst M. de, Tromp, E., Schilt, M., Krol, L., Gersonius, B., Vlaming, C., Valkenburg, L. & Peeters, R. 2009. Waterrobuust bouwen. De kracht van kwetsbaarheid in een duurzaam ontwerp. KvR report 016/2010. Beter Bouw- en Woonrijp Maken/SBR, Rotterdam. ISBN 978-90-8815-017-3.

Van de Ven, F.H.M., Tjallingii, S.P., Baan, P.J.A., Eijk, P.J. van, Rijsberman, M. 2006 Improving urban water management planning. Proc. International Symposium on Lowland Technology 2006, Saga, Japan.

Van den Brink, M., Meijerink, S., Termeer, C., & Gupta, J. 2013. Climate-proof planning for flood-prone areas: assessing the adaptive capacity of planning institutions in the Netherlands. Regional Environmental Change, 1-15.

Van Den Hurk, B., Klein Tank, A., Lenderink, G., Van Ulden, A., Van Oldenborgh, G. J., Katsman, C., & Drijfhout, S. 2006. KNMI climate change scenarios 2006 for the Netherlands. De Bilt: KNMI.

Van der Brugge, R. 2009. Transition dynamics in social ecological systems. Rotterdam.

Van der Brugge R., Rotmans J. & Loorbach D. 2005. The transition in Dutch water management. Regional Environmental Change 5, 113–135.

Van der Brugge, R., Rotmans, J., 2007. Towards transition management of European water resources. Water Resources Management 21, 249–267.

Van Herk, S., Herder, P. M., Jong, W. M. de & Alma, D. S. M. 2006. Innovative Contracting in Infrastructure Design and Maintenance. International journal of critical infrastructures (ijcis), (ISSN 1475-3219), 2 (2/3), 187-200.

Van Herk, S. 2008. Urban Flood Management Dordrecht

Van Herk, S., Zevenbergen, C., Rijke, J. & Ashley, R. 2009. Collaborative research to support transition towards integrating flood risk management in urban development, Submitted to International Conference on Urban Flood Management, Paris.

Van Herk, S., Zevenbergen, C., Rijke, J. & Ashley, R., 2011a. Learning and Action Alliances for the integration of flood risk management into urban planning: a new framework from empirical evidence from The Netherlands. Environ. Sci. Policy 14, 543-554

Van Herk, S., Zevenbergen, C., Rijke, J. & Ashley, R. 2011b. Collaborative research to support transition towards integrating flood risk management in urban development, Journal of Flood Risk Management, Volume 4, Issue 4, December 2011, Pages: 306–317

Van Herk, S., Kelder, E.T.G., Bax, J.H., Van Son, E., Waals, H., Zevenbergen, C., Stone, K.,Gersonius, B. 2011c. Gebiedspilot meerlaagsveiligheid Eiland van Dordrecht; Tussenrapportage ter inspiratie.

Van Herk, S., Rijke, J., Zevenbergen, C., Ashley, R., 2012a. Governance of integrated flood risk management to deliver large scale investment programmes: delivery focused social learning in the Netherlands., Floodrisk 2012 - 2nd European Conference on flood risk management, Rotterdam, Netherlands.

Van Herk, S., Rijke, J., Zevenbergen, C., Ashley, R., 2012b. Transition in governance of River Basin Management in the Netherlands through social learning, International Sustainability Transitions Conference, Copenhagen, Denmark.

Van Herk, S., Kelder, E.T.G., Bax, J.H., Gersonius, B en Hulsebosch, M. 2012c. Meekoppelmethode voor het klimaatbestendig maken van bestaand stedelijk gebied, Pilot: Wielwijk Klimaatbestendig, Concept rapportage.

Van Herk, S., Zevenbergen, C., Gersonius, B., Waals H., Kelder, E.T.G. (2013a) Process design and management for integrated flood risk management: exploring the Multi Level Safety approach for Dordrecht, the Netherlands. Netherlands. In press at Journal of Water and Climate Change

Van Herk, S., Rijke, J., Zevenbergen, C., Ashley, R., (2013b) Understanding the transition to integrated flood risk management in the Netherlands. Journal of Environmental Innovations and Societal Transitions (EIST). doi: 10.1016/j.eist.2013.11.001

Van Herk, S., Rijke, J., Zevenbergen, C., Ashley, R., Besseling, B. 2013c. Adaptive multi-level governance through social learning: River Basin Management in the Netherlands. Earth System Governance, Tokyo, Japan.

Van Herk, S., Rijke, J., Zevenbergen, C., Ashley, R., Besseling, B. (in press) Adaptive co-management and network learning in programmes; case study Room for the River. Journal of Environmental Planning & Management. doi: 10.1080/09640568.2013.873364

Van Herk, S., Rijke, J., Zevenbergen, C., Ashley, R., (under review) Attributes for integrated Flood Risk Management projects; case study Room for the River. Under review at: International journal of River Basin Management

Van Raak, R. 2010. The transition (management) perspective on long-term change in healthcare. 3 In: Transitions in Health Systems: Dealing with Persistent Problems J. E. W. Broerse and J. F. G. Bunders. Amsterdam, VU University Press

Van Ree, C. C. D. F., Heilemann, K., Morris, M. W., Royet, P. and Zevenbergen, C. 2011. FloodProBE: technologies for improved safety of the built environment in relation to flood events. Environ. Sci. Policy, doi:10.1016/j.envsci.2011.03.010.

Van Twist, M., Ten Heuvelhof, E., Kort, M., Olde Wolbers, M. Van den Berg, C. and Bressers, N. 2011. Tussenevaluatie PKB Ruimte voor de Rivier.

Veerbeek W., Ashley R M.. Zevenbergen C., Rijke J. and Gersonius B. 2010. Building Adaptive Capacity For Flood Proofing In Urban Areas Through Synergistic Interventions. First International Conference on Sustainable Urbanization (ICSU 2010). Hong Kong, China, 15-17 December 2010.

Veerbeek, W, and Zevenbergen C. 2009. Deconstructing urban flood damages: increasing the expressiveness of flood damage models combining a high level of detail with a broad attribute set, Journal of Flood Risk Management 1, 45-57

Veerbeek, W., Ashley, R., Zevenbergen, C., Gersonius, B., Rijke, J. 2012. Building adaptive capacity for flood proofing in urban areas through synergistic interventions. 7th International Conference on Water Sensitive Urban Design, Melbourne, Australia.

Verhagen J., Butterworth J., Morris M. 2008. Learning alliances for integrated and sustainable innovations in urban water management. Proc. 33rd WEDC International conference, Accra, Ghana.

Von Malmborg, F. 2007. Stimulating learning and innovation in networks for regional sustainable development: the role of local authorities. Journal of Cleaner Production, 15, 17, 1730-1741

Voß, J.P. 2007 Innovation processes in governance: the development of'emissions trading'as a new policy instrument. Science and Public Policy 34, 329-343.

Voß, J.P., Bornemann, B. 2011 The Politics of Reflexive Governance: Challenges for Designing Adaptive Management and Transition Management. Ecology and Society 16, 9.

Walker W. 2000. Entrapment in large technology systems: institutional commitment and power relations. Research Policy 29, 833–846

WaterForum Online, 2012. Interview met Han Vrijling, hoogleraar constructieve waterbouw TU Delft. http://www.waterforum.net/Interviews/2783-hoogleraar-vrijlingkeuze-politiek-voor-meerlaagse-veiligheid-bizar#sthash.baAEONTe.dpuf [accessed 28/04/12]

Waterman, R. E. 2008. Integrated Coastal Policy Via Building with Nature: Towards an Integrated Coastal Policy Via Building with Nature, Using Forces and Materials Present in Nature. Opmeer Drukkerij bv: The Hague

Watson, C. C., Biedenharn, D. S., & Wohl, E. E. 2011. Comparison of flood management strategies. Inland flood hazards: human, riparian, and aquatic communities, 381-393.

Watson, D. and Adams, M. 2010. Flood-Resistant Design for Sites and Communities, in Design for Flooding: Architecture, Landscape, and Urban Design for Resilience to Flooding and Climate Change, John Wiley & Sons, Inc., Hoboken, NJ, USA. doi: 10.1002/9781118259870.ch10

Webb, E.J., Campbell, D.T., Schwartz R.D., & Sechrest, L. 1966. Unobtrusive Measures: Non-reactive Research in the Social Sciences. Chicago, USA.

Wenger, E. 2000. Communities of practice and social learning systems. Organization 7, 225-246.

Wheater H., Evans E. 2009. Land use, water management and future flood risk. Land Use Policy, 26S, S251-S264.

White I 2010. Water and the City: Risk, Resilience and planning for a sustainable future. Routledge, London and New York. ISBN13: 978-0-415-55333-9

White I. 2008. The absorbent city: urban form and flood risk management. Proc. ICE Urban Design and Planning 161, December 2008, Issue DP4, 151–161.

Whyte, W.F. 1989. Advancing Scientific Knowledge Participatory trough Action Research. Sociological Forum 4, 367-385

Williams, M. 2001. Problems of knowledge: A critical introduction to epistemology (pp. 32-37). Oxford: Oxford University Press.

Wolsink, M. 2006. River basin approach and integrated water management: Governance pitfalls for the Dutch Space-Water-Adjustment Management Principle. Geoforum, 37, 473-487.

Woltjer, J, & Al, N. 2007. The integration of water management and spatial planning; journal of the American Planning Association 73, 211 – 22.

Wong, T.H.F. & Brown, R., 2009. The water sensitive city: principles for practice. Water Science and Technology 60, 673–682.

Yin, R.K. 2003. Case study research: design and methods, SAGE publications, Applied Social Research Methods, volume 5

Yovel, E. 2013. Shaping Resilience: Mainstreaming Disaster Risk Reduction into Land Use Planning. Draft Working Paper. UN/ISDR: Incheon, Korea

Zevenbergen, C., Veerbeek, W., Gersonius, B., Van Herk, S. 2008. Challenges in urban flood management: travelling across spatial and temporal scales. Journal of Flood Risk Management, 1, 81-88

Zevenbergen, C., Cashman, A., Evelpidou, N., Pasche, E., Garvin, S., & Ashley, R. (2010). Urban flood management. CRC Press Inc..

Zevenbergen, C., Rijke, J., van Herk, S., Ludy, J., Ashley, R. 2013a. Room for the River: International relevance. Water Governance 02/2013

Zevenbergen, C., van Herk, S., Rijke, J., Kabat, P., Bloemen, P., Ashley, R., Speers, A., Gersonius, B., Veerbeek, W. 2013b. Taming global flood disasters. Lessons learned from Dutch experience. Natural Hazards, 65(3), 1217-1225

Zhou Q., Quitzau M-B., Hoffmann B., Arnbjerg-Nielsen K. 2013. Towards Adaptive Urban Water Management: Up-Scaling Local Projects. International Journal of Architecture, Engineering and Construction 1, 63-72

Bibliography

Journal papers

Rijke, J., Vessels Smith, J., Gersonius, B, Van Herk, S.; Pathirana, A., Ashley R., Wong, T., Zevenbergen, C. (subm). Operationalising resilience to floods and drought across multiple layers of intervention: protection, prevention and preparedness. Submitted to Journal of Hydrology

Koukoui, N., Gersonius, B., Schot, P.P., Van Herk, S. (under review) Adaptation Tipping Points and Opportunities for the Management of Urban Flood Risk - Case Study in Dordrecht, the Netherlands. Under review at Journal of Water and Climate Change.

Van Herk, S., Rijke, J., Zevenbergen, C., Ashley, R., (under review) Attributes for integrated Flood Risk Management projects; case study Room for the River. Under review at: International journal of River Basin Management

Van Herk, S., Rijke, J., Zevenbergen, C., Ashley, R., Besseling, B. (in press) Adaptive co-management and network learning in programmes; case study Room for the River. Journal of Environmental Planning & Management. doi: 10.1080/09640568.2013.873364

Van Herk, S., Rijke, J., Zevenbergen, C., Ashley, R., (2013) Understanding the transition of integrated flood risk management in the Netherlands. Journal of Environmental Innovations and Societal Transitions (EIST). doi: 10.1016/j.eist.2013.11.001.

Rijke, J., Brown, R., van Herk, S., Zevenbergen, C. (under review) Governance for strategic planning and delivery of adaptation. *Global Environmental Change.*

Rijke, J., Van Herk, S., Zevenbergen, C., Ashley, R., Hertogh, M., Ten Heuvelhof, E.F. (in press) Programme success through aligning project and programme management. In press at International Journal of Project Management.

Zevenbergen, C., Rijke, J., van Herk, S., Ludy, J., Ashley, R. (2013) Room for the River: International relevance. Water Governance 02/2013

Zevenbergen, C., van Herk, S., Rijke, J., Kabat, P., Bloemen, P., Ashley, R., Speers, A., Gersonius, B., Veerbeek, W. (2013). Taming global flood disasters. Lessons learned from Dutch experience. Natural Hazards, 65(3), 1217-1225.

Rijke, J., Van Herk, S., Zevenbergen, C., Ashley, R. (2012) Room for the River: delivering integrated river basin management in the Netherlands. International journal of River Basin Management, Volume 10, Issue 4.

Van Herk, S., Kelder, E.T.G., Zevenbergen, C., Meijers, E.., Waals H., Van Os, N., Segeren, T., Hendriks, M. (under review). Samenwerken aan beleidsverandering in waterbeheer; Impact, organisatie en perspectieven op de Learning & Action Alliance. Submitted to Bestuurskunde

Van Herk, S., Zevenbergen, C., Gersonius, B., Waals H., Kelder, E.T.G. (2013) Process design and management for integrated flood risk management: exploring the Multi Level Safety approach for Dordrecht, the Netherlands.

Netherlands. Journal of Water and Climate Change. doi: 10.2166/wcc.2013.171

Rijke, J., Brown, R., Zevenbergen, C., Ashley, R., Farrelly, M., Van Herk, S., Morison, P., (2012) Fit-for-purpose governance: A framework to operationalise adaptive governance. Environ. Sci. Policy. Volume 22, October 2012, Pages 73–84

Van Herk, S., Zevenbergen, C., Rijke, J., Ashley, R., (2011) Learning and Action Alliances for the integration of flood risk management into urban planning: a new framework from empirical evidence from The Netherlands. *Environ. Sci. Policy*, 14, 543-554

Van Herk, S., Zevenbergen, C., Rijke, J., Ashley, R. (2011). Collaborative research to support transition towards integrating flood risk management in urban development, Journal of Flood Risk Management, Volume 4, Issue 4, December 2011, Pages: 306–317

Kelder, E.T.G. Van Herk, S., Minnema, P.A. (2010). Waterveiligheidsbeleid. Meerlaagsveiligheid: van theorie naar praktijk. RBOH nr. 32

Zevenbergen, C., Veerbeek, W., Gersonius, B. and Van Herk, S. (2008). Challenges in Urban Flood Management: travelling across spatial and temporal scales, Journal of Flood Risk Management 1:2(81-88)

Herk, S van, Herder, P M, Jong, W M de & Alma, D S M (extern) (2006). Innovative Contracting in Infrastructure Design and Maintenance. International journal of critical infrastructures (ijcis), (ISSN 1475-3219), 2(2/3), pp. 187-200

Conference papers

Koukoui Gersonius Schot Van Herk (2013) Adaptation Tipping Points for the Management of Urban Flood Risk - Case Study in Dordrecht, the Netherlands. European Climate Change Adaptation Conference (ECCA), Hamburg, Germany.

Dudley E., Richard Ashley, Natasa Manojlovic, Sebastiaan van Herk, John Blanksby (2013) Learning and Action Alliances for innovation and active learning in a European context. International Conference on Flood Resilience. Exeter, UK.

Van Herk, S., Rijke, J., Zevenbergen, C., Ashley, R., Besseling, B. (2013) Adaptive multi-level governance through social learning: River Basin Management in the Netherlands. Earth System Governance, Tokyo, Japan.

Van Herk, S., Rijke, J., Zevenbergen, C., Ashley, R., (2012) Governance of integrated flood risk management to deliver large scale investment programmes: delivery focused social learning in the Netherlands., Floodrisk 2012 - 2nd European Conference on flood risk management, Rotterdam, Netherlands.

Van Herk, S., Rijke, J., Zevenbergen, C., Ashley, R., (2012) Transition in governance of River Basin Management in the Netherlands through social learning, International Sustainability Transitions Conference, Copenhagen, Denmark.

Van Herk, S. (2012) Challenges in Policy & Practice: Governance of integrated planning & Flood Risk Management. Submitted to 12th International Convention of Melaka Twin Cities 2012: be prepared for emergency disaster preparedness and risk reduction, Malaysia.

Waals, H., Kelder, E.T.G., Van Os, N., Van Herk, S. (2011), Working apart together – filling the water safety gap. Submitted to 5th International Conference on Flood Management, Japan.

Rijke, J., Anema, K., Zevenbergen, C., Van de Ven, F., Van Herk, S. (2011) Capacity building dissected: what to learn in order to mainstream sustainable urban water management in the Netherlands? In proceeding of: 2nd International Conference on Sustainability Transitions, 2nd International Conference on Sustainability Transitions, Lund, Sweden

Van Herk, S., Zevenbergen, C., Rijke, J., Ashley, R. (2009). Collaborative research to support transition towards integrating flood risk management in urban development, Submitted to International Conference on Urban Flood Management, Paris.

Zevenbergen, C., Veerbeek, W., Gersonius, B., Thepen, J., and van Herk, S. (2008) Adapting to climate change: using urban renewal in managing long-term flood risk. in: Flood Recovery, Innovation and Response. D. Proverbs, C.A. Brebbia and E. Penning-Rowsell (eds), WIT Transactions on Ecology and the Environment, Volume 118, ISBN: 978-1-84564-132-0.

W. Veerbeek, C. Zevenbergen, Van Herk, S. (2008) Urban Flood Management: Towards a Flood Resilient Urban Environment; In proceeding of: SIBICO International (eds) Conference Proceedings Water resources systems management under extreme condition, Moscow, Russia

Van Herk, S., Herder, P. M., de Jong, M. J., & Alma, D. (2004, October). Opportunities and threats for granting more design space to road contractors. In Systems, Man and Cybernetics, 2004 IEEE International Conference on (Vol. 5, pp. 4643-4649). IEEE.

Book chapters

Van Herk, S.: Contributions to Chapter 12, Capacity building & governance. In: Zevenbergen, C., Cashman, A., Evelpidou, N., Pasche, E., Garvin, S., & Ashley, R. (2010). *Urban flood management.* CRC Press/Balkema – Taylor & Francis Group, London.

Zevenbergen C, Gersonius B, Puyan N, van Herk S (2007) Economic feasibility study of flood proofing domestic dwellings. In: Advances in Urban Flood Management. R Ashley, S Garvin, E Pasche, A Vassilopoulos, and C Zevenbergen (Eds.). Taylor and Francis. ISBN: 978 0 415 43662 5.

Gersonius, B., Zevenbergen, C. and Herk, van S. (2007): Managing flood risk in the urban environment: linking spatial planning, risk assessment, communication and policy. In: Adaptive and Integrated Water Management: coping with complexity and uncertainty. C Pahl Wostl, P Kabat, and J Möltgen (Eds.). Springer ISBN 978 3 540 75940 9.

Reports

Bastiaanssen, R., van Herk, S., Mekel, B. (2012), MARE WP4 Policy Development, Overview of activities and impact of the MARE project in 4 Learning and Action Alliances.

Van Herk, S., Kelder, E.T.G., Bax, J.H., Gersonius, B en Hulsebosch, M. (2012), Meekoppelmethode voor het klimaatbestendig maken van bestaand stedelijk gebied, Pilot: Wielwijk Klimaatbestendig, Concept rapportage.

Van Herk, S., Kelder, E.T.G., Bax, J.H., Van Son, E., Waals, H., Zevenbergen, C., Stone, K.,Gersonius, B. (2011), Gebiedspilot meerlaagsveiligheid Eiland van Dordrecht; Tussenrapportage ter inspiratie.

Stalenberg, B., Vreeker, R., Edelenbos, J., Verweij, S., Winsemius H.C., Van Herk, S., Smid, G.J.P., Van Middendorp, B. (2011) Klimaat in Ruimtelijke Keuzes; het dialoogondersteunend afwegingskader (DAK); theorie en praktijk.

Building the Netherlands Climate Proof: Urban Areas; a contribution to the study "Opportunities for climate adaptation in the Netherlands" of the Netherlands Environmental Assessment Agency. 1201082-000-VEB-003. © Deltares, 2010.

Van Herk, S. (2008). Urban Flood Management Dordrecht

Acknowledgements

My gratitude goes to my supervisor Prof. Chris Zevenbergen for convincing me to embark on this adventure, motivating and guiding me at the most difficult moments and helping me to enjoy all others. Your refreshing vision, opportunistic approach and relentless optimism were the compass for my research in an emerging domain. Thanks to my supervisor Prof. Richard Ashley for mentoring me, continuously advocating, engaging and empowering us young researchers and improving every single publication, often on instant demand.

This journey would have been inconceivable without my colleagues at the Flood Resilience Group. When we co-founded the FRG, William Veerbeek and Berry Gersonius envisioned the fun and value of science and the complementarity of our work. Your enthusiasm, together with that of Chris, Richard and Ellen Brandenburg, was decisive for me to join the FRG alongside my 'normal job'. Jeroen Rijke, my dear colleague, I don't know if or when I would have finished my PhD without you. You have been my loyal and reliable companion. Thank you for your flexibility and empathy, especially considering my limited presence. Thanks for providing peer pressure to keep me going and thanks for hours of brainstorming and reflection. The FRG family grew and the synergies remained. I am grateful to all other FRG colleagues, Kim Anema, Taneha Bacchin, Assela Pathirana, Koen Olthuis, Natasa Manojlovic, to provide the evidence that together we can accomplish more, also individually. Thank you Chris for creating the conditions for such an environment of support, cross-fertilisation and fun.

This research would not have been possible without the support of Bax & Willems Consulting Venturing, my employer. Thank you Laszlo Bax and Menno van Rijn for allowing me the freedom and flexibility to develop a scientific career in parallel to our consultancy work. I am especially grateful to Rolf Bastiaanssen. With your hard work and versatility you serviced our clients extremely well and created space for my PhD. My ideas have been enriched by your critical and constructive reflections on case studies and the applicability of theoretical frameworks. Special thanks are due to Laurianne van der Plas who helped me with a remarkable work-ethos to finish the thesis. I want to acknowledge all my colleagues for their patience and empathy for my fluctuating availability.

Many thanks go to the practitioners, policy-makers, funders and scientists I worked with in the research projects and case studies. Together we have been able to make a difference to practice, whilst enriching science. The combination has been highly motivating. Special mention goes to Ellen Kelder, all participants of the LAA Dordrecht and all colleagues at Dura Vermeer. You continuously sought innovation, experimenting with new approaches to combine learning with action, exploring new visions on IFRM policy and engaging ever more stakeholders and disciplines. I am grateful to Bert Satijn and Corné Nijburg who created and managed the Living with Water programme. You funded and believed in my work that benefited greatly from your programme. I owe a huge acknowledgement to all professionals involved in the Room for the River (RftR) programme. You were remarkably open to share your experiences, both positive and negative, for the benefit of future programmes and science. As a Dutch citizen you make me proud of your work. Thank you Liesbeth van Riet Paap, Ben Broens and Ingwer de Boer for giving us the opportunity to evaluate RftR and make us part of the community you shaped. Thank you all for showing me the value of your various perspectives and disciplines.

The quality of this research has greatly benefited from the feedback of peer-reviewers, the Scientific Board members and many fellow researchers (e.g. COST C22). The professionals I met during assignments and conferences around the world provided new perspectives and helped me to frame my own work.

Finally I have the opportunity to immortalize my gratitude to my family and friends for the happiness they bring to my life. Also to reach this milestone, your love and support have been of pivotal importance. Vanessa, thank you for having endured and having stimulated me whilst I was traveling or working long evenings, weekends and holidays on my thesis. You bring joy and love to my life in so many ways. Papa, mama, and Nathalie, thank you for unconditionally loving me for who I am and stimulating me to live life in my own way and to the fullest. Especially during my visits you were flexible to enjoy time together and provided your loving care to make my travels possible.

This research has been done with pleasure. And this has been the nicest part to write. During my PhD I have learnt to help and ask for help, to give and receive, to inspire and be inspired, and to say thank you. We can't do it alone. Not a PhD, not life, not IFRM.

Professional biography

Sebastiaan van Herk holds an MSc degree (cum laude / with honours) in systems engineering, policy analysis and management from Delft University of Technology (2003), studied industrial engineering at the UPC Barcelona and economics at the Erasmus University in Rotterdam. Mr. Van Herk works for Bax & Willems Consulting Venturing (Spain) since 2004 as a strategy consultant in Open Innovation, in 2007 he was appointed member of B&W's management team and in 2012 as partner. He has defined, executed and coordinated Open Innovation, Research & Development, Business Development and policy projects for various industrial multinational and SME companies; governmental bodies at different administrative levels; and research institutes, including several European and Global collaborative projects. Mr. van Herk mainly works in the domains of: environment; water; and construction, but has occasionally delivered projects in energy, aeronautics, textiles, ICT broadening his experience in effective collaboration early in his career.

In 2008 he started a (part-time) PhD at UNESCO-IHE and Delft University of Technology on governance, learning and collaboration in Integrated Flood Risk Management and Spatial Planning. The PhD research combines several individual research projects and related scientific publications. He is a founding member and researcher at the Flood Resilience Group.

He has been appointed UNISDR (United Nations International Strategy for Disaster Reduction) campaign Advocate for Resilient Cities. Sebastiaan also contributes to educational; training; and capacity building programmes. He is an experienced speaker and facilitator of workshops at international conferences (approximately 4-6 per year).

Previously Mr. van Herk worked 2 years in R&D and NBD departments in the construction industry at HILTI AG (Liechtenstein) and Dura Vermeer Group (Holland) respectively.

Sebastiaan van Herk speaks English, Spanish, German, Dutch and Catalan fluently and has notions of French.

For Product Safety Concerns and Information please contact our EU representative GPSR@taylorandfrancis.com Taylor & Francis Verlag GmbH, Kaufingerstraße 24, 80331 München, Germany

Batch number: 10722120

Printed by Printforce, the Netherlands